HUMAN EVOLUTION AND SURVIVAL

Candid Conversations on the Human Story

Edited by Ellis E. McDowell-Loudan

State University of New York, Cortland

cognella
San Diego, CA

Bassim Hamadeh, CEO and Publisher
Christopher Foster, General Vice President
Michael Simpson, Vice President of Acquisitions
Jessica Knott, Managing Editor
Kevin Fahey, Cognella Marketing Manager
Jess Busch, Senior Graphic Designer
John Remington, Acquisitions Editor
Jamie Giganti, Project Editor
Brian Fahey, Licensing Associate
Kate McKellar, Interior Designer

First published in the United States of America in 2013 by Cognella, Inc.

ISBN: 978-1-5165-5126-2

cognella™

www.cognella.com 800.200.3908

CONTENTS

SECTION THREE: ETHICS AND CAREERS: HUMAN ISSUES OF TODAY

DEDICATION

To my husband Gary, first and foremost, and to the generations of students who have injected ideas into the challenging study of what it means to be human.

ACKNOWLEDGMENTS

Thanks to the encouragement and enthusiastic assistance of numerous people at Cognella Publishing, it has been a pleasure to develop this book. John Remington (Senior Field Acquisitions Editor, Northeast Regional Office—Connecticut) was my first contact at Cognella. From the start, he convinced me that this would be a project that was worthwhile and that I would enjoy. He was right. Kevin Fahey (Marketing Manager) sent me some formidable questionnaires to fill out and utilize as background for part of what I was doing. Not only did these help focus portions of the work, but Kevin's assorted e-mails were useful on several occasions. Jess Busch (Senior Graphic Designer) is amazing. I knew what I wanted for a cover for the book, somewhere in the recesses of my mind. Sure enough, she managed to excavate my mind and create something I was envisioning. Then Kevin e-mailed me to ask whether I would like a subtitle for the book. We e-mailed some back and forth, and the result was one of his original ones: "Candid Conversations on the Human Story." Jess quickly provided an amended cover page with the subtitle in red—it really popped. Meanwhile, throughout these exchanges, there was some delightful communication with Jamie Giganti, my project editor. What a wonderful, talented individual she is! Jamie has helped polish all the rough spots in my chapters. She provided invaluable advice and clarified things that I didn't understand correctly. Jamie, you are a miracle worker and thank goodness you have the eagle eye for editorial issues of all kinds. Thus, it is evident that the Cognella team is a joy to work with, and I thank each and every one of them.

The resources from the Cognella library of publisher partner content (http://library.universityreaders.com) were a pleasure to explore and provided many opportunities for potential inclusion in the text. Thanks for this type of resource; it is a clever approach for the preparation of a volume like this one.

It is a distinct pleasure to include a new article by my colleague, Nasser Malit, Assistant Professor of Anthropology at SUNY Potsdam. His interpretation of Neandertals and modern hominins is an exciting addition to this volume.

Finally, my husband, Gary, photographed, critiqued, suggested changes, helped with choices, encouraged, and generally participated in much of the preparation of the book. He reminds me always that there is more to our lives together than the book.

INTRODUCTION

By Ellis E. McDowell-Loudan

The title chosen for this collection of articles is *Human Evolution and Survival*. The challenge has been to try to pick useful articles to excite, incite, stimulate, infuriate, and generate critical thinking among an unknown group of readers. Choices have been arbitrary and reflect the interests of the author of this collection.

How was the book title chosen? It is the title of the course for which it was developed, taught by the author, at the State University of New York at Cortland College. This set of readings, based on ideas generated by ideas from students and colleagues, seems useful for the second decade of the twenty-first century.

WHAT DO HUMAN EVOLUTION AND SURVIVAL MEAN?

By Ellis E. McDowell-Loudan

What does HUMAN mean? Dictionaries, including the Internet's *Wikipedia*, provide an assortment of definitions. Biology textbooks offer still more definitions, as do illustrations in art books, where geometric form or muscle structure may be emphasized. Religious and philosophical texts introduce concepts of spirit and soul for an entirely different approach to humankind. It is possible to list, discuss, and debate the meanings at great length, depending upon what contexts are appropriate for the specific use of the term. As much as possible, a biological emphasis and what is referred to as an anthropological approach will be our strategy.

Humans are ... ? Every once in a while, somebody's parent says: "Stop acting like an animal!" The irony, of course, is that biogenetically, we *are* animals. That's not what the parent was referring to, though. He or she referred to something completely different, as you and I understand quite clearly. The exclamation was applied to behavior (never my behavior, of course) that was socially inappropriate for a specific time and place.

This usage of the term "animal" had nothing to do with the criteria Carolus Linnaeus used in his *Systema Naturae*, published in 1758, to classify the creatures he placed in the Animal Kingdom. Thus, if a human is an animal, as Linnaeus demonstrated and as has been documented thoroughly since that time, there must be a lot more involved in defining the term.

There are a lot of variations on that theme. Dogs, gorillas, elephants, lions, jellyfish, flatworms, oysters, lobsters, birds, and fish, among others, are not difficult to differentiate from us—at least if we are doing the distinguishing! Each of these creatures appears to recognize its own kind and its own specific environmental place, or niche. The frustrated parental statement about animal behavior, though, was not the time for us to respond with scientific information.

To continue with our definition, humans are able to do a particular set of things in a certain combination of ways, which differ from the way other animals do these things. At the same time, other animals are able to do things that we cannot do. For instance, without an oxygen tank, you and I cannot survive underwater for long periods of time the way a whale can. A whale requires a vast amount of sea water and space, as well as the opportunity to come up for air periodically. It is an animal and a mammal, just as we are.

Each animal is unique and each has both biogenetically advantageous and disadvantageous traits. As environments change over time, some individual members of each species (whether it is a plant or animal species) have characteristics which help them survive at that time and place, or not. The changing environments and individuals within them interact in such a way that some members of the groups are better able to reproduce offspring successfully than other members of the groups at that time and place. To grasp the HUMAN part of our study, we're going to go off in several directions. One will look at genetics to try to start at the beginning of the beginning, with cells—or even earlier than that, with the best understanding we have of the chemical reactions that came from outside our own universe! These clues will be present in the genetics articles.

A second direction we shall take (and I hesitate to mention the term because some of us grew up claiming to dislike "history") will be to look at ideas from the past. Some of them may seem odd when we look at them now. These clues will turn up in numerous articles. However, if some of the suggestions made by students in recent classes are valid, the contexts of the bits of history may stimulate discussion, debate, and if nothing else, disgust!

Today, we have access to information from people from all over the world. You can look up almost anything you want or need to know in minutes! However, there are abundant contradictions in the information available. Critical evaluation of all you hear and read is essential.

Who could believe that the Earth is flat or that you could fall off the edge if you sail out to sea too far? Who could think that every living thing is absolutely fixed and never, ever changes at all from one generation to another? Of course, if the priest or other major leader told you these things were so and you were taught to trust and believe him or her absolutely, this would become truth for you.

There were many teachings that were passed from generation to generation of human beings and some continue to be, among some of us. In many cases, if you did not live near the sea or if a ship went out to sea and never came back to port, it might appear true, especially if you were warned over and over that those who sailed out to the horizon and never returned had fallen off the edge. In the centuries before the 1600s–1700s, very few people in Europe had access to any formal education other than what was taught by the local village priest. Even this would have been meager for most children due to the work schedules of farming communities and the selective nature by which family members would be sent to school. Travel of any major distances overland was infrequent for most as well. Few could read and write.

Even in more recent times, formal education has been unequally distributed around the globe and continues to be variable in many parts of the United States as well. Ironically, we often pay more attention to the educational limitations of those in other parts of the world than we do in our own backyard.

When we stop to look at various myths and misconceptions in our own parts of the world, we are often surprised that they aren't true! Many of these are very slow to disappear because once we learn a particular thing, it sticks—even something that might seem foolish when we stop to think about it. For instance, "the bigger the brain, the smarter you are," (Einstein had a small brain) or "put butter on a burn" (we use butter for frying, not cooling).

Some religious groups insist that those who follow a path of scientific study that recognizes connections among all living things cannot possibly have a belief in God or any religion. This is not necessarily so. Many scientists are religious individuals and many religious leaders have strongly developed scientific backgrounds. There are many ways to understand science and the scientific method as one part of our world, while one's beliefs and faith are something distinct from this.

A third direction of our travel is one of both humor and sadness. We shall look at some of our closest relatives, the great apes: chimpanzees, bonobos, orangutans, and gorillas. Many of us have enjoyed watching them in zoos, sometimes without recognizing they are prisoners in cages where people are looking at them and laughing at their "antics." Why are they so "cute and funny?" We recognize many of our own behaviors in what they do, don't we?

Surprise! In the wild, the great apes live in diverse kinds of family groups, eat different diets, and spend a great deal of time doing things differently from what we see in the zoo. Not only that, but in most cases, these "folks" may never have seen their home territory, may have been born in captivity and were human-nursery—raised, because their mothers don't know how to nurture them properly, since probably they, too were born in captivity. Not only that, but many are endangered species susceptible to human diseases. As a result, we can't risk losing any of their offspring.

Through the laboratory and zoo studies, supported by some of the wonderful work done by researchers like Jane Goodall (chimpanzees), Dian Fossey , Michele Goldsmith, and George Schaller (gorillas), and Birute Galdikas (orangutans), we have background information and much more concern for the care of these animals in captivity.

Efforts to rebuild environments where nonhuman primates will have homelands that resemble their native ones are part of the effort for all of us to "be green." It is to our mutual advantage to try to rebuild our world as much as we can to reclaim some of what we are losing before it is too late for all of us. A brief section on primates will provide a taste of this rich area of research. Perhaps as important will be the sections on other environmental issues which are applicable to all parts of the world and all of us (see "Death of an Icon" and "Hope Amid the Ruins").

A fourth direction we shall travel is the ongoing story of the fossil record. Paleoanthropological discoveries of fossils possibly relating to our human family are reported all the time! When a discovery occurs, announcements with pictures zip around the world. Instantly, there are questions and comments, with numerous interpretations. Of course, some may be negative and others erroneous. This is a disadvantage of the ease with which we are able to share knowledge. However, the benefits tend to outweigh disadvantages, and this is why some understanding of what it all means is of value.

What is EVOLUTION? Again, this concept has been used and abused for many generations and continues to cause problems. Evolution is change through time.

In the context in which we are using it, the term refers to change through time of any and all living things. It is an ongoing process, a continuum. Time is part of this same concept and is part of the understanding of the term continuum.

What is SURVIVAL? This term in our usage differs from some of the other uses you will see elsewhere. We are not discussing "survival gear for back-to-school," "survival foods for final exam week," and all those other gimmicks that are part of the advertising world.

Survival in our use has only one important purpose: successful reproduction for a species. Indeed, this is the definition of any species: Individuals who reproduce offspring that are capable of reproducing offspring that are capable of reproducing successfully. This definition may sound glib or redundant. Remember, our main interest is the evolution of ourselves as the human species and our survival as a species.

Therefore, when we embark on our examination of human beings and their/our survival and reproductive success, there is another vital phenomenon which is essential for our survival: culture!

CRITICAL THINKING DISCUSSION

1. Is it enough to look only at the biological aspects of the survival concept when studying Human Evolution and Survival? Why or why not?
2. Are biogenetic aspects of the human species the only determinants for all types of learned behavior? Why or why not?
3. What four directions are taken here to study ourselves? Is one more important than the others?
4. What are some examples of cultural baggage? What are some that you know of that are not listed here?
5. If you were to select one point from this chapter to write about, which one would it be and why?

Human culture is a configuration or set of learned patterns of behavior shared by human beings. It is modified continuously and transmitted from person to person and group to group and from generation to generation through language and all our other learned behavior patterns. Thus, what we are studying is human biocultural evolution and survival.

Section One

GENETICS ISSUES

THE GOOD BOOK

Help! Is There An Evolutionary Biologist in the House?

By Kenneth W. Krause

Struggles to defend proven yet popularly controversial science from the forces of religious excess are all too familiar to readers of this column. But how many of us realize that even our highly trained and trusted physicians have received little or no instruction in evolutionary biology? Currently, no U.S. medical school offers a course on evolutionary medicine, and not one maintains a department of evolutionary biology or has a single faculty member specializing in evolutionary perspectives on healthcare.

Meanwhile, valuable evolutionary applications to medicine are being discovered at a dizzying rate. That many doctors fail to appreciate evolution as the central organizing principle of all biology, some experts contend, is an embarrassment to the profession and a dangerous disservice to the patients who depend on it. So, as awe-inspiring advances in genetics, genomics, biotechnology, and neuroscience mount, biologists around the world are stepping forward to insist that the fateful moment has arrived for medicine to fully embrace its scientific bedrock.

On April 2-3, 2009, the National Academy of Sciences in Washington, DC, held an Arthur M. Sackler Colloquium titled, "Evolution in Health and Medicine." According to organizer Stephen Stearns, Yale University ecologist and evolutionary biologist, the objective was to "raise awareness in the medical community that evolutionary biology had helpful insights into medical problems that were being missed by other perspectives." The event's provocative results were released in a multi-paper supplement to the January 26, 2010, *Proceedings of the National Academy of Sciences.*

"Resistance," Stearns confided to me, "is both practical and political." Medical curricula are already packed with essential or at least contributive classes, of course, and are "thought to be functioning well by those teaching it." Yet evidence from the front lines seems to betray some

Kenneth W. Krause, "Help! Is There an Evolutionary Biologist in the House?" *The Humanist,* vol. 70, no. 4, pp. 39-41.

glaring deficiencies. For example, Stearns noted, "The over-prescription of antibiotics, particularly by pediatricians, often for viral infections that do not respond to antibiotics, continues to be widespread despite the fact that the rapid evolution of antibiotic resistance is the best understood example of evolutionary medicine." The consequences of such uninformed treatment to patients, he warned, could prove "difficult, expensive, and perhaps insoluble."

Co-organizer Randolph Nesse, University of Michigan professor of psychiatry, considers the gap in evolutionary education among physicians "wide and serious." By the time they enter practice, he told me, most physicians have received far more instruction on quantum physics than in evolutionary biology, and because they lack a solid theoretical foundation, many have no idea why people age or even how cancer arises. It's akin to "engineers not understanding the first law of thermodynamics," he says, "or thinking that heavier objects fall faster than lighter ones."

In the summary *PNAS* article, "Making Evolutionary Biology a Basic Science for Medicine," an international team of thirteen medical and biological experts, including Nesse and Stearns, argued for more integrative medical education and offered specific suggestions. A more penetrating comprehension of evolutionary processes, they argued, "helps to correct the prevalent dependency on the metaphor of the body as a designed machine." To the contrary, the human form is more accurately characterized as a "jury-rigged system" better understood as a series of compromises shaped by natural selection to maximize reproductive viability rather than health. Only when medical students are required to correct the former, erroneous mindset, the authors insist, will they gain a "deeper understanding of the body and why it is vulnerable to disease."

Peter Gluckman, Alan Beedle, and Mark Hanson, biomedical experts and authors of a new textbook titled *Principles of Evolutionary Medicine* (Oxford, 2009), appear to agree. Contrary to their patients' best interests, they say, physicians generally settle into "reductionist approaches" to medicine, focusing narrowly on body parts and systems and the immediate or "proximate" causes of disease. "Ultimate" or evolutionary causes are largely ignored. But the latter perspective, the authors propose, "not only helps to identify research questions, but also allows engagement with individual patients in ways which promote understanding of their current health status and which also contribute to the design of appropriate interventions in public health."

Principles offers physicians (and their concerned patients) eight possible pathways by which to conceive of and address a given medical condition from an evolutionary perspective. The most common and perhaps easily understood mechanism results from conspicuous "mismatches" between our slowly evolving genomes and our rapidly shifting cultural circumstances. A wide variety of common but often serious pathologies—everything from lactose intolerance and myopia to obesity and addiction—are thought to arise from these clashes.

Gluckman emphasizes infant formula feeding, a cultural innovation promoted most ardently in the 1970s, as a particularly notorious example. Nutritionally distinct from breast milk, formula is said to have caused obesity, reduced cognitive ability, and even increased infant mortality among children in the developed world. In Israel, the authors note, infants were left brain damaged or brain dead because their formulas lacked the thiamine, or vitamin B1, their mothers might otherwise have provided.

In addition to the patient's course of illness, the authors urge physicians to consider two much deeper historical levels when evaluating a problem. First, developmental history from the point of conception can reveal how plasticity and disruption have influenced the patient's phenotype. Second, evolutionary history—since the inception of hominins, *Homo sapiens,* agriculture, and perhaps the patient's specific ancestral or cultural lineage, can deliver genetic and other more intuitive clues revealing altered risk levels and potential treatments. A more "holistic approach," they conclude, "will pay enormous dividends for medicine and amply repay the effort needed to adopt it."

Similarly, in the *Textbook of Evolutionary Psychiatry: The Origins of Psychopathology* (Oxford, 2008), Martin Brüne, professor of psychiatry at the University of Bochum, Germany, demands "radical revision" of his profession's current paradigm characterizing knowledge of evolutionary principles as interesting but superfluous. "If psychiatry wants to survive as a medical discipline," he counsels, "it can no longer be satisfied with a knowledge base that covers just 50 percent of what is needed for a full comprehension of psychopathological conditions." Understanding of the ultimate causes of cognition, emotion, and behavior, he believes, is "absolutely essential" to the intelligent diagnosis and treatment of mental disorders.

According to Brüne, psychiatrists frequently presume that proximate factors like early development and neurochemistry can fully explain mental pathology. But a more complete investigation of any given behavior, he urges, will also include an attempt to trace its evolutionary roots and an analysis of its potential adaptive value. Though once selected for some benefit, psychopathological symptoms often reflect "extremes of variation" that have become maladaptive, or "dysfunctional," because of their inappropriate frequency, intensity, or context.

Take fear, for example—clearly an ancient and highly adaptive trait. When expressed under circumstances posing no objective threat, however, fear becomes pathological anxiety. But when we ask the appropriate evolutionary question—here, why it might be "cheaper" for a person in actual danger to be more rather than less frightened than necessary—we can see why it makes good selective sense to lower the threshold of release. This nimble evolutionary induction, Brüne proposes, might explain why so many psychopathological conditions involve anxiety.

Though primarily crafted for academic settings, both Gluckman and Brüne are accessible to frequent readers of this column. Utterly gorged with cutting-edge medical research and fascinating insight into the broader human condition, these generalized yet meticulous texts will leave readers feeling invested and eager for a great deal more—which is amply provided by either of two brilliant, multi- author, multi-topic anthologies.

In *Evolutionary Medicine and Health: New Perspectives* (Oxford, 2008), biological anthropologists Wenda Trevathan (New Mexico State), E. O. Smith (Emory), and James McKenna (Notre Dame) have assembled twenty-three original essays contributed by forty-four experts from Australia, France, Norway, Switzerland, the U.K., and the United States. Highlighting diet and nutrition, sex and reproduction, and social disorders and chronic diseases, the editors have judiciously chosen to address specific health concerns of critical interest to the public at large.

Which is not to imply that more complicated medical issues were ignored. For example, in "Genes, Geographic Ancestry, and Disease Susceptibility," Douglas Crews and Linda Gerber, public health experts from The Ohio State and Cornell Universities, respectively, discuss three means of determining disease risk through personalized medicine—self-reported race and ethnicity, geographic ancestry, and individual profiles. They also compare each model's medical utility, and, of course, explain how the relevant debate should be guided by evolutionary principles.

Though commonly exploited by modern clinicians and researchers, "race," the authors confirm, is merely a social construct that "does not reflect any specific biological reality." As such, its therapeutic merit is severely limited. And why would any medical professional armed with even the most basic working knowledge of natural history suppose otherwise? As Crews and Gerber point out, we have no reason to believe that disease-promoting genes have been subjected to the same evolutionary pressures as those for skin color or hair texture, or that such alleles are in any way linked.

Methods drawing upon geographic ancestry are much preferred. Scientists have discovered many genes that predispose specific populations to diabetes, hypertension, and obesity, for example. But while a few such alleles are indeed unique to those groups, the authors caution, "most are shared to some degree across populations." Thus, the most accurate reflectors of natural selection and genetic drift are individual DNA profiles.

"The molecular revolution," Crews and Gerber summarize, "has provided new and powerful tools for profiling one's genetic risks [for chronic degenerative conditions] and for understanding gene-environment and gene-culture interactions predisposing individuals, families, and groups to specific conditions." Unfortunately, the revolution's clinical potential has yet to be seriously developed—perhaps in part because its scientific foundation remains somewhat obscure to so many clinicians and researchers.

An equally diverse suite was collected by Stearns and Jacob Koella, biologist at the Imperial College London, in the second edition of *Evolution in Health and Disease* (Oxford, 2008). Although evolution can't provide easy answers to every medical dilemma, say Stearns, Nesse, and Harvard biologist David Haig in the text's opening and introductory chapter, evolutionary thinking can provide precious insight and even "save lives" when physicians prescribe antibiotics, administer vaccinations, advise couples having difficulty conceiving, and treat cancer, among many other things.

Patients might want their doctors to know, for example, why Western women experience demonstrably higher rates of cancer than women living without contraception in pre- industrial cultures. More menses, of course, mean more cell divisions and a greater likelihood of dangerous mutations. Physicians and researchers need to recognize, as the authors point out, that "there is nothing biologically normal about the regular monthly period." Developing contraceptives that don't induce periods but do provide the appropriate level of estrogen for maintaining bone strength and avoiding osteoporosis may be the best solution.

And perhaps all pediatricians should recognize the value of worms. Having evolved in human guts to down-regulate our immune responses, certain parasites have also reduced child susceptibility to autoimmune diseases by flagging the production of antibodies. But because modern environments have become so antiseptic, autoimmune diseases have made dramatic comebacks as worm infestations have waned. This rather peculiar evolutionary mismatch, Stearns, Nesse, and Haig instruct, "helps to explain the current epidemics of asthma, type I diabetes, and even leukemia." But certain informed doctors, they note, have actually made headway by treating children with parasitic fauna.

Of course scientists have no desire to completely overhaul medical school curricula. Instead they suggest modest yet substantive contributions where evolutionary material would be clearly helpful. A free, intelligent, and continuing dialogue between biology and medicine would no doubt benefit both disciplines. But well-informed patients inclined to take responsibility for their own health and fitness will be the biggest winners. Inherently predictive, evolutionary thinking emphasizes prevention over treatment and, as such, personal empowerment over dependence or helplessness.

You can learn more at the *Evolution and Medicine Review* (http://evmedreview.com). The next "Good Book" will further explore evolutionary applications to medicine in two critical contexts—women's health and the now global epidemic of obesity.

OH NO!

Those So-Called Mendelian Traits Aren't!

By Ellis E. McDowell-Loudan

It is fortunate that science education moves forward and is a two-way process. It is also lucky that sometimes while perusing the Internet for material to supplement a lecture, contradictions catch the eye. That is what happened in May 2012. John H. McDonald's "Myths of Human Genetics" website became apparent when I saw the title and illustrations provided (http://udel.edu/~mcdonald/index.html). To my dismay—indeed—horror, Professor McDonald listed and discredited all the examples I'd been using in a laboratory for classes for years! They include examples taken from a basic genetics handbook—one that was utilized by many faculty as a simplistic way to convey basic Mendelian concepts.

These concepts included visible and fascinating special genetic traits about each of us. The problem is that their means of inheritance is more complex than the "either dominant or recessive" criteria suggested. They are not simple Mendelian traits. First, we don't necessarily have just a single dominant clockwise growth hair whorl pattern or a single recessive counterclockwise growth hair whorl pattern, for instance.

There is a number of alternative alleles (forms of each gene) at each locus. The set of traits isn't that simple! In classes, we casually noted that there were variations and commented that mutations probably accounted for this, but ignored the issue.

Second, there appear to be other genes on the hereditary strands of DNA, which are adding to what each trait may produce, so that in at least some cases, a trait may not be Mendelian at all. It may be polygenic—that is, it may be affected by more than one gene, additively. This "now you tell me" example even extends to one of our more common textbook examples, the ABO blood type system. The situation may exist where it appears that an offspring of parents who show the codominant AB blood type has the O blood type! What's wrong here?

At the "Stanford at the Tech Understanding Genetics 2012" website, an article by Dr. Barry Starr explained that even this Mendelian trait has a rare allele called cis-AB. According to Starr, "about 0.03% or 3 out of every 10,000 Koreans have this blood type ... about 0.001% of Japanese have the trait, one out of every 100,000"(http://genetics.thetech.org/ask/ask427). The individual who carried this trait would have type AB blood in the phenotype, since A and B are codominant. The genotype would be cis-AB. The other parent would be AO or BO or OO, so that the recessive form would appear in the offspring.

A third revision to the way in which we perceive human polymorphism results from what we know today about how some of Mendel's genes work. Mendel noted some of the basic connections between "the color of various plant components. Specifically, Mendel noticed that plants with colored seed coats always had colored flowers and colored leaf axils. (Axils are the parts of the plant that attach leaves to stems.) Mendel also *observed* that pea plants with colorless seed coats always had white flowers and no pigmentation on their axils. In other words, in Mendel's pea plants, seed coat color was always associated with specific flower and axil colors [Lobo, 1 (2008) Pleiotropy: One gene can affect multiple traits. Nature Education 1(1)]." Pleiotropy is a fancy word for this. One gene or locus influences what happens with other traits. It may trigger growth, eye color, or other attributes. Illustrations for this are shown at http://www.biog1105-1106.org/demo/106/unit06/7.pleitropy.html), Cornell University.

Drat! I am disgusted that the warning signs that should have jumped out at me years ago did not do so. Every semester, students have enjoyed this lab, laughing at each other as we have checked out our diversity. We discussed the range of variation in the way in which each of us displayed the traits—including diversity of the degree of a trait, absence, or presence in an extreme form of it. We even speculated about possible alternative alleles, which would explain why we were not exactly one way or another in our appearance of each trait!

We contemplated the idea of environmental factors that might have influenced such variation or possible mutations which had added new alleles (forms) of the traits. However, we continued to utilize the same Mendelian trait approach to the laboratory.

The degree of variation should have been a red flag. It should have sent the curious off to the library. It should have sent me to the Biology Department to ask questions. Might it have sent an enterprising student to their Biology professor to find out whether they were getting erroneous information from their Anthropology professor? (No, I shall not try to pass the buck ...)

I have not been corrected by any of the genetics majors taking my classes, though, despite my fervent requests that they do so if they encounter an error in what I say or what our text appears to present. On occasion, we have had some discussions on topics or items that do need revisions. This is an essential part of the true educational process, from my point of view.

However, this embarrassing discovery that the entire laboratory was based on something make-believe, just because a researcher thought it was simpler than the truth does not make any sense to me. Why teach something incorrect as a model to illustrate a concept? We make enough mistakes without conscious ones! To my previous students, I hope you have discovered the truth and the numerous resources available to us for studying and learning about human diversity. The array of possible traits, the amount of variation on the theme of each of these genes we thought were so simple, is much more amazing and logical than the former view that these alleles never mutated, even though all sorts of other ones did.

1. Below are illustrations of hair whorl pattern for comparison with your own, that of fellow students, and maybe family members. Clockwise growth—dominant? Counterclockwise—recessive? But have you one or two whorls? Why? The online sources discuss some of the

explanations. Example A is the counterclockwise pattern (a clock with a second hand is used to help in case you aren't clear what "clockwise" means in today's world of digital clocks). The extra line in the scalp is a scar, not a natural spare hair part. Example B shows a somewhat ambiguous clockwise whorl pattern and a second counterclockwise whorl farther down the scalp. Long hair and wavy hair like this woman's often conceal the hair whorl patterns, as do various hair treatments, of course. Example C has a clockwise hair whorl. This example also presents the issue of thinning hair, just as thick or long hair and fine hair make the definition of this trait ambiguous on occasion. Tight curls are extremely confusing, too. What's the implication? The genetics of hair whorl patterns, now that they have been studied extensively and scientifically, are not based exclusively on a Mendelian genetic trait. They are not directly associated with a person's left-handedness or right-handedness, despite what was believed from observations in informal settings (http://udel.edu/~mcdonald/hairwhorlpatterns/html).

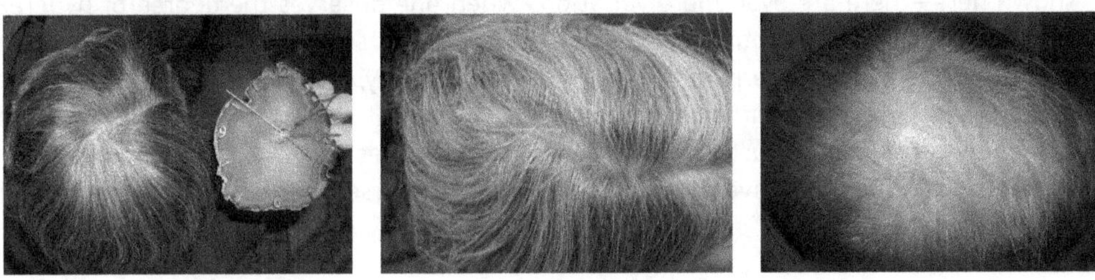

A. B. C.

2. Four examples: B shows "Darwin's point" clearly (right), a projection of cartilage on the ear's helix (edge). A (left) has an indentation which is slightly firm, but doesn't really show the "point" any more—dominant, or absence of this—recessive. Some have a large or smaller one—or two—Value? C and D (bottom left and right), no Darwin's point.

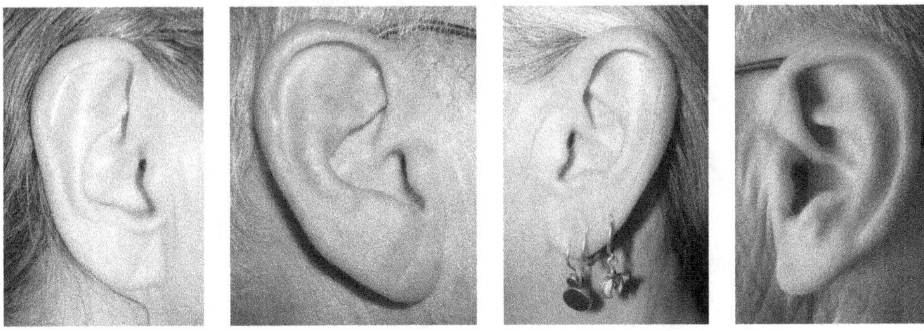

A. B. C. D.

3. The same ear examples show attached earlobe (left A and slightly detached, C)—recessive and unattached earlobe (right B and more so D)—dominant. Value? Significance?
4. The fists show the mid-digitals, the smaller of the fists in A have remnants of hairs, now worn off from aging (and much archaeological labor, perhaps?). Neither of the fists in B indicates the presence of hair, despite intensive gardening, carpentry, and other work with the hands. Hair—on the mid-digit—dominant; absence of hair—recessive. Lots of variation

with this one, too—one or more fingers may exhibit the trait. There appear to be multiple variants which may imply geographic and environmental explanations. Value?

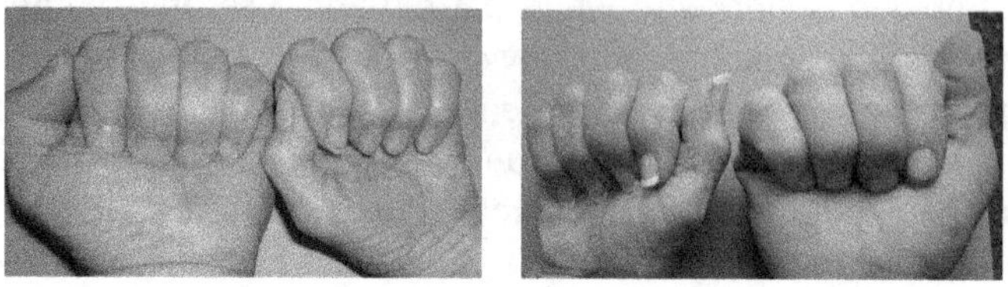

A. B.

5. Hitchhiker's thumb—**inability to hyperextend**, as shown in A—dominant. Hyperextensible, at an angle of 50 degrees or greater—recessive. Note the range of variability stated and shown here—there are even more variations when one observes the degree of hyperextensibility in action. And other features of the hands—possibly interrelated? Pleiotropy? Polygenic ones? Possible environmental factors? Certainly, A and C with varied degrees of inability to hyperextend differ markedly in shape (as well as sex), and the striking variations between B and D illustrate other aspects of the recessive trait. In each instance, the four hands and fingers have marked dexterity and diverse abilities not relating to this attribute.

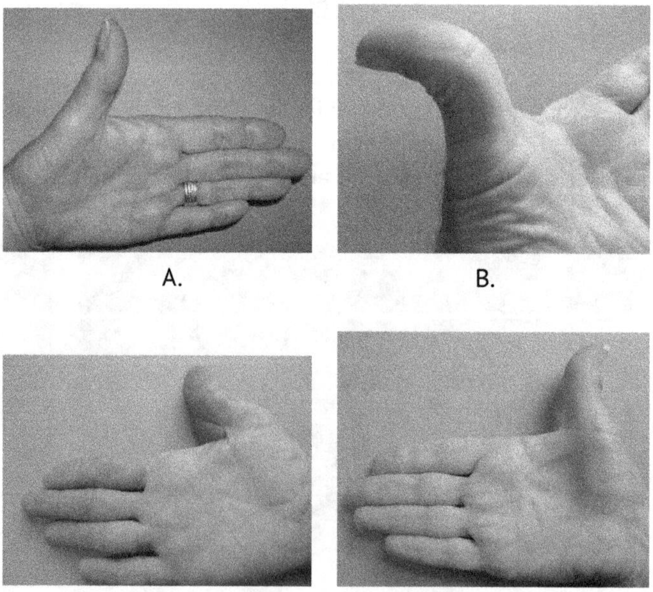

A. B.

C. D.

Fingerprints are another of the fascinating attributes we hear about as traits that are unique to the individual. They offer identifiable features used to trace lost children, document individuals for passports or various licenses, and assist with criminal investigations. Techniques for fingerprint analysis have become more and more sophisticated, although inked marks on paper still do the job adequately in many cases.

Certainly, the use of inked prints for many generations and the numerous examples of prints on surfaces which would lend themselves to comparison with inked prints suggest that these records continue to play a role in today's world. Below are suggestions of the uses for such identifications. These fingerprints bring back the sorrow and horror of the John F. Kennedy assassination.

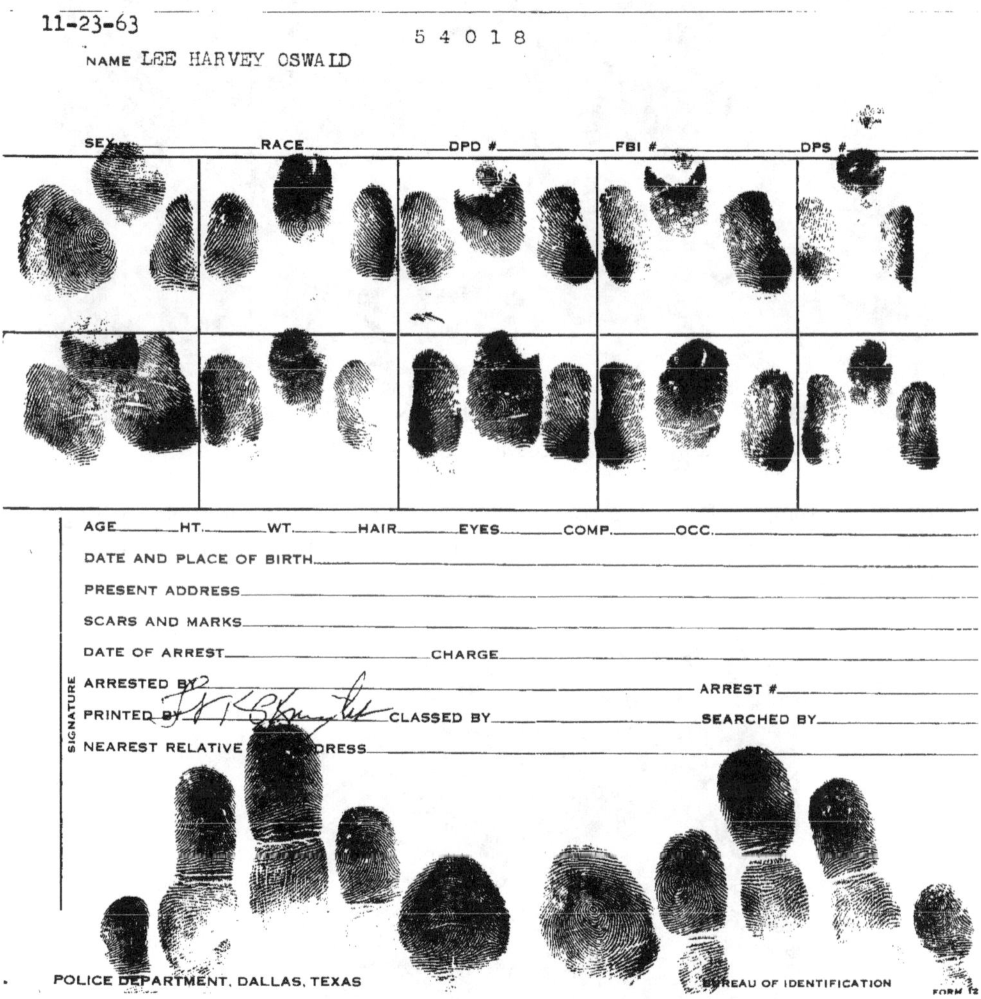

11-23-63

54018

NAME LEE HARVEY OSWALD

SEX RACE DPD # FBI # DPS #

AGE HT. WT. HAIR EYES COMP. OCC.

DATE AND PLACE OF BIRTH

PRESENT ADDRESS

SCARS AND MARKS

DATE OF ARREST CHARGE

SIGNATURE

ARRESTED BY ARREST #

PRINTED BY CLASSED BY SEARCHED BY

NEAREST RELATIVE ADDRESS

POLICE DEPARTMENT, DALLAS, TEXAS

BUREAU OF IDENTIFICATION

FORM 12

Source: Dallas Office of the City Secretary. Copyright in the Public Domain.

Similarly, if the location for gathering fingerprints lacks sophisticated equipment for obtaining the prints, the traditional inking approach continues to be worth the effort. Today, it is relatively easy to transfer these data to the new technology through computerized scanning systems.

Source: www.navy.mil/view_single.asp?id=62255. Copyright in the Public Domain.

Below is an example of a laptop computer that verifies the computer user by fingerprint (2010). There are similar forms for index fingers in use by the military to identify individuals at various checkpoints today.

Source: "Reading Fingerprint" © Avenafatua, (CC BY-SA 3.0) at: http://commons.wikimedia.org/wiki/File:Reading_Fingerprint.jpg.

Note: I want to thank Gary L. Loudan, Mardis Kelsen Schuhle, and John Schuhle for permitting use of their heads and hands for this chapter, and Gary for his excellent photographs of these illustrations as well.

Chapter 4 provides an example of the use of fingerprinting on a national level for identification of all people within India. This process appears to be one which would enable everyone to receive equal access to medical aid. It sounds like an ideal approach to a problem which is found throughout the world—unequal access to resources.

MAKING THE POOR COUNT

India's National Identity Scheme

By Keshava Guha

Biometric national identity systems are rare in industrial democracies. Recent attempts to introduce them in the United Kingdom and the United States have been defeated by civil libertarian opponents. In both cases, the proposed identity systems were closely linked to national security and the US-led Global war on Terror. A biometric identity database is now being introduced on an unprecedented scale in the world's largest democracy: India. By 2015, 600 million Indians will be issued a Unique Identification number (UID), a number linked to a resident's fingerprints, iris scan, and facial photograph. This time the primary motive is not national security, but rather social benefits.

The UID began life in a form similar to that of the abortive UK and US systems. In 2003, the then-ruling center-right coalition proposed the creation of a Multipurpose National Identity Card for the purpose of fighting terrorism and crime. After the coalition failed to win reelection, its successor, the Congress-led United Progressive alliance (UPA), temporarily suspended the idea. The UPA's 2009 election manifesto, however, revived the notion of a biometric identity system but with a new function. The Unique ID, later branded "Aadhaar," is intended to revolutionize the nature of welfare schemes in India.

Aadhaar is the brainchild of the Unique Identification Authority of India, a new government agency headed by Nandan Nilekani, billionaire software entrepreneur and former Chairman of Infosys Technologies. The scheme has the backing of the three most important figures in the Indian government: Prime Minister Manmohan Singh, UPA Chairperson Sonia Gandhi, and her son Rahul, and the first numbers will be issued before the end of 2010. It took the United States three decades to create a de facto identity system in the form of the Social Security number; India is attempting to do the same in biometric form in under five years.

Keshava Guha, "Making the Poor Count: India's National Identity Scheme," *Harvard International Review*, vol. 32, no. 3, pp. 8-9.

That OECD countries do not have biometric identity schemes is a frequent criticism of Aadhaar. Yet its advocates point out that the absence of identity is a unique impediment to India's poorest, who are thereby excluded from the very welfare programs that are created to help them. For example, millions of the poorest Indians do not possess ration cards, as they are unable to pay the requisite bribe to the local *panchayat* (village council).

Two schemes that could be cleaned up in the short term are the Public Distribution System (PDS), which distributes food grains, and the National Rural Employment Guarantee Scheme (NREGS), which provides enrollees with 100 days work at slightly more than US$2 per day. It is estimated that of the US$9 billion annual outlay for NREGS, up to US$3.6 billion is lost to mid-level corruption; enrollees consequently typically receive only 60 percent of the promised income.

By giving such concrete identity credentials to the poorest Indians, many of whom have never possessed any kind of identity proof in their lives, Aadhaar can ensure that PDS food grains reach their intended source. The benefits to NREGS will come through Aadhaar's pledge of financial inclusion. It is proposed that anyone with an Aadhaar number will receive a basic, no-frills bank account; in a country where only a small fraction of the population use formal financial services today, this change would have a revolutionary impact. Enrollees in NREGS could have their salary directly deposited to their biometrically-authenticated bank account, entirely bypassing the myriad network of middlemen that account for the 40 percent income loss.

In the long-run, financial inclusion could be an essential step in a complete overhaul of India's welfare state. For decades, many economists have recommended that India replace its labyrinthine system of subsidies and price controls, many of which benefit the wealthy, with a system of direct cash transfers to the poor. Bank accounts for all will make this possible. This possibility, however, has never been formally acknowledged by the ruling Congress Parties and is bitterly opposed by the left.

Aadhaar neither confers citizenship nor inherently guarantees benefits; rather, it provides an unimpeachable identity with which citizens will be able to avail themselves of the benefits to which they are entitled. Residents of India will not, at the moment, be required to apply for an Aadhaar number; rather, as with the US Social Security number, enrollment will be technically voluntary. Service providers will, however, be allowed to mandate that customers possess a number.

While Aadhaar has received a remarkably enthusiastic reception from its key prospective beneficiaries—the poor—it has also attracted considerable criticism from the kinds of civil libertarians that successfully derailed ID schemes in other countries. Opponents of Aadhaar assert that it is part of a hidden agenda of the creation of a surveillance state, an unacceptable transformation of the relationship between the citizen and the state. They point to the creation of a National Intelligence Grid (NATGRID, a proposed national security database), and argue that a biometric number will be misused to track citizens' personal data.

The UIDAI has responded to these concerns by asserting that Aadhaar will only require the most basic demographic information of its enrollees and that it is not linked to NATGRID or any other government database. As of September 2010, it was unclear whether or not the Ministry of Home Affairs intended to proceed with NATGRID. In any case, the success or failure of Aadhaar, which is the subject of legislation currently pending before the Lok Sabha, India's lower house of parliament, will be judged on its effectiveness in using the proof of identity to deliver benefits to the hundreds of millions of deserving Indians who have been denied them for decades.

Section Two

PRIMATE EVOLUTION AND SURVIVAL

GENES, BRAINS, AND CULTURE

From Monkey to Human

By Jean-Pierre Changeux

It is undeniable that the brain of a monkey or of an ape is not the brain of a human being. Many anatomical and behavioral traits distinguish apes from men including the use of language and access to consciousness. It is equally undeniable, in the genomic era, that such species-specific differences are bound to DNA-encoded mechanisms. Such genetic mechanisms would establish limits, create an "envelope," of the prenatal but also of the post-natal development of the brain and of its cognitive predispositions. We are thus at a dramatic turning point in biological thinking. From the gene sequences stored, *in silico,* in computer memory, all the protein molecules that make up our brain or contribute to its development and plasticity are known or should soon be known. Ultimately, we should be able, in principle, to compute the main features of the organization of our brain from the knowledge of our genetic endowment. At least, we should be able to identify the "signature" of our human nature in terms of DNA sequences and to specify which DNA elements make the difference between the human brain and the monkey brain.

As stated by Karl Zilles, the human brain is more than simply a large monkey (or ape) brain (figure 5.1). In the course of evolution, the mean brain size increases. In about 2.6-3 million years in human lineage the brain volume increases approximately from 457 cm^3 in *Australopithecus africanus* to 1,355 cm^3 in *Homo sapiens,* with intermediate values of 552 cm^3 for *H. habilis* and 1,016 cm^3 for *H. erectus* (Carroll, 2003). In this absolute scale, the macaque brain looks rather tiny with its 63 cm^3, while gorillas and chimpanzees approach the direct human ancestors with respectively 348 and 305 cm^3 (Semendeferi et al., 1997). But if one compares on an allometric scale the different brain compartments by reference to a hypothetical insectivore of the same size, some particular structures of the brain become disproportionately enlarged in the course

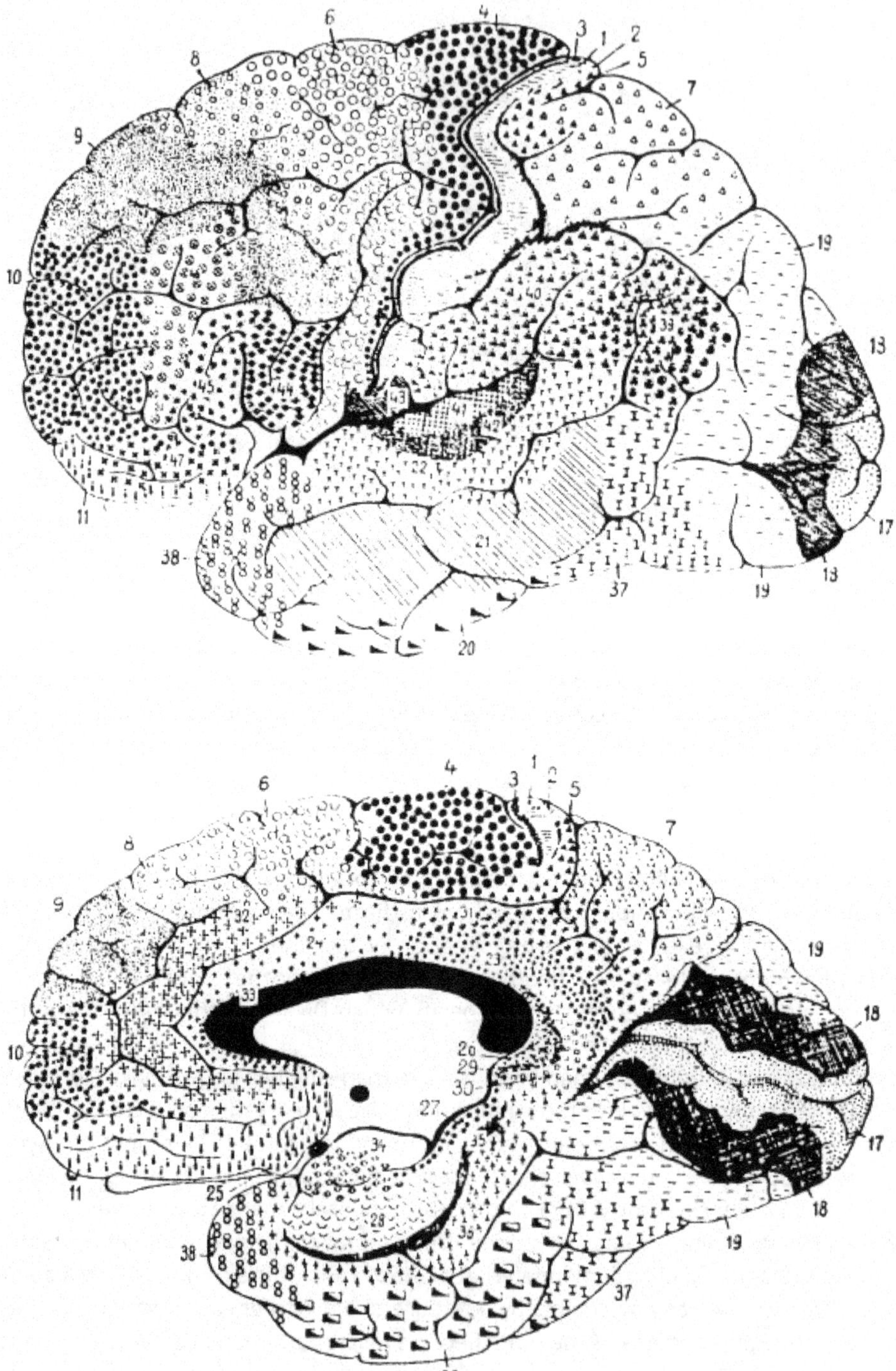

Figure 5.1 Comparison of the maps of the cerebral cortex areas of the human *(top)* and monkey *(bottom)* as published by Brodmann in 1909.

of brain evolution (see chapter 3, this volume). The paleocortex does not differentially increase from insectivores to humans. On the other hand, the neocortex expands most intensively, largely as a consequence of the exceptional increase of the multimodal association cortex. The number of cortical areas progresses from 10 to 20 in lower mammals (specialized for instance for vision, audition, motor control, and so forth) up to a brain with a very high relative cortical surface in humans, with as many as 100 cortical areas (including association ones) integrated within multiple parallel and hierarchical levels (see Mountcastle, 1998). There is a long-standing notion that the prefrontal cortex, which is mobilized by "higher" cognitive functions, is disproportionately larger in humans (Brodmann, 1912). Recent volumetric analyses by magnetic resonance, however, show a more modest though still significant relative increase of frontal cortex relative volume from 30 percent in the macaque, to 35.4 percent in chimpanzees and 37.7 percent in humans (Semendeferi et al., 2002). One area of the prefrontal cortex, area 10 specifically involved in cognitive functions, is, however, relatively enlarged and more specialized in humans relative to apes (Semendeferi et al., 2001).

Apes and monkeys do not produce and understand speech. Yet, monkeys have been assumed to possess the equivalent of Broca and Wernicke language areas, although without the rich connectivity that characterizes language processing in humans (see Rauschecker, Tian, & Hauser, 1995; Aboitiz & Garcia, 1997; Deacon, 1997). Moreover, a precursor of the human left-right asymmetry of these areas has been found by magnetic resonance imaging (Cantalupo & Hopkins, 2001; Gannon, Holloway, Broadfield, & Brown, 1998). At a more microscopic scale, the cytoarchitecture of the cerebral neocortex has a common six-layered structure in primates, including humans, but varies significantly from one area to the other and from species to species with an unexpected diversity of neurotransmitter receptors patterns. Thus an interesting paradox arises. On the one hand, the monkey and human brains share a common plan of organization, yet striking species-specific regional differences unambiguously distinguish them.

Are we in a position to account altogether for these common and different features on the basis of our current knowledge of human genome? This is a challenging issue for the molecular biologist since the number of genetic events which took place in the past million years to achieve such remarkable morphogenetic changes looks astonishingly small!

GENES AND EVOLUTION

Before entering the discussion about genes and brains, we should be clear about what we mean by gene. The definition began with the experiments of Mendel on peas and the observations that the inheritance of traits—form, color of the seeds, and so on—follows mathematical rules. This implies the stability of the traits over generations and thus the occurrence of, then, rather enigmatic hereditary "factors." The term *gene* appears in the early 1900s as an abstract concept to account for these hereditary factors, which were, soon after, associated with specific regions of chromosomes by Morgan and his collaborators. Subsequently, Beadle restricted the definition of the gene to the nature of its product: a protein, thus the famous statement "one gene—one enzyme." Progress in recombinant DNA technologies ultimately lead to a new definition of the gene that is now referred to as the "complete chromosomal segment for making a *functional* product" (Snyder & Gerstein, 2003). The definition includes the expression of a product, which has to be functional, as well as regulatory regions (like promoters) in addition to the protein-coding ones. At the DNA sequence level, structural genes can be identified by start and stop codons that bound a so-called open reading frame within which transcription may take place. Incidentally, the human genome contains a significant contingent of "dead genes" or pseudogenes, which contain frame shift or stop codons in the middle of coding regions, thus preventing them from giving a functional product. Demonstration that a gene is functional requires additional evidence that is not given by straightforward sequence inspection: it is, for instance, the direct demonstration of transcription into mRNA in situ as well as functional inactivation by mutation. Yet, most of the attempts to "annotate" genes in the sequenced genomes do not always satisfy all these criteria simultaneously. It should be emphasized that analyzing the raw nucleotide sequences gives only provisory assignments that often have (or will have) to be revised.

Evaluations of the total number of genes from fully sequenced genomes have nevertheless been published. In man, the current estimate is 30,000–31,000 genes, which are thought to occupy a little more than 1.5 percent of the whole genome (IHGSC, 2001; Venter et al., 2001). Most of the DNA sequence in our chromosomes is thus noncoding. This does not mean, as we shall see, that these noncoding repetitive or intervening sequences have no function. Moreover, proteomics—the study of the proteins actually expressed by a given genome—has to deal with the fact that each individual gene may express as its product more than one protein species. First, it may give several splice variants mRNA and proteins; moreover, post-translational modification may create an additional polymorphism (there are about 100 such modifications known). According to O'Donovan et al. (2001) the number of different protein molecules expressed by the human genome is probably closer to a million than to the accepted number of genes, around 30,000. This number is still very small compared to the 50–100 billion neurons in the human brain, and the statements "one gene—one neuron" or even "one gene—one synapse" are blatantly false. There is thus, in sheer number, a striking limitation, a *parsimony,* of available genetic information, not only to code for the body but also, more dramatically, to code for the brain (Changeux, 1983). In order to face this striking parsimony of genetic information, combinatorial mechanisms in gene activities have to be postulated. Today we have access to the full genomic sequences of several major organisms in addition to man (30,000–31,000 genes), for instance: the yeast *Saccharomyces cerevisiae* (6,144 genes) (Goffeau et al., 1996), the plant *Arabidopsis thaliana* (25,706 genes) (Arabidopsis Genome Initiative, 2001), the worm *Caenorhabditis elegans* (18,266 genes) *(C. elegans* Sequencing Consortium, 1998), the fruitfly *(Drosophila melanogaster)* (13,338 genes) (Adams et al., 2000), the fugu fish (Aparicio et al., 2002) and the mouse (27,000–30,500 genes) (Mouse Genomic Sequencing Consortium, 2002). This basic information offers the opportunity to compare the full genetic endowment of far distant species and, in principle, to identify the gene sequences that direct the building up of the cells, their assembly into a multicellular organism, and, more specifically, the development of the central nervous system.

A first simple and naive reasoning is that the evolution of the number of genes parallels the evolution of the complexity of these organisms and, in particular, of their brain. In fact, this is not the case. First of all, the "core proteome," the number of nonredundant proteins coded by the genome of multicellular organisms such as the fly (8,065 proteins) and the worm (9,453 proteins) is only twice that of the yeast single cell (Goffeau et al., 1996; *C. elegans* Sequencing Consortium, 1998; Adams et al., 2000). About 30 percent of the protein coding genes are shared by fly and worm, and nearly 20 percent by fly and yeast. About 1,308 groups of protein coding genes are shared between humans, fly, worm, and yeast, representing respectively 3,129, 1,445, 1,503, and 1,441 protein species (IHGSC, 2001). The shared proteins referred to as *housekeeping* exhibit functions that are common to all eucaryotes: they mediate, in particular, DNA replication and repair, biosynthesis, folding, and degradation of proteins as well as their transport and secretion. These household proteins are essential for cell life, but they do not make the difference between yeast, the fly, and human beings.

Close comparison of the human genome data with those of the invertebrates reveals that only 7 percent of the gene families are unique to vertebrates, and only 12 percent of them are thought to be concerned by the brain. Homologies of many of the genes whose alterations are associated with human neurological diseases, are already found in *Drosophila* (Rubin et al., 2000). The genes that predispose, for instance, to Tay-Sach disease, Duchenne muscular dystrophy, lissencephaly, or fragile X mental retardation possess homologs in the fly genome. Those that cause amyotrophic lateral sclerosis or adenoleukodystrophy are even present in the yeast. Only a few of them look specific to humans (or vertebrates). They are, for instance, those which cause Charcot-Marie-Tooth or Creutzfeld-Jacob diseases. The fraction of genes that, in the human genome, look proper to humans remains unexpectedly small.

Among the protein coding genes that distinguish vertebrates from invertebrates are the genes engaged in inflammation and immunity, hemostasis, development, cell death, and nervous system function and organization. From the worm and the fly to the human genome, there is thus an increase in the sheer number of members of protein coding gene families that are selectively involved in the development of the nervous system and in neural signalling (IHGSC, 2001; Venter et al., 2001). In particular, as we will see, the genes coding for transcription factors increase dramatically, but so do those involved in the formation of connections, axon guidance (like the ephrins and their receptors), adhesion molecules (such as the proteoglycans), or nerve growth (NGF) or trophic factors (neuregulins). In addition, the genes coding for proteins taking part in the cytoskeleton (like actins and spectrins) expand together with the proteins that compose the myelin sheet. In other words, there is a marked expansion of the proteins involved in building up the neuronal axonal and dendritic arborizations and their interconnections. This is also the case for molecules involved in the propagation of electrical (ion channels) and chemical signals (the opioid peptides are present only in vertebrates). Still, an unexpected nonlinearity (Changeux, 1983) persists in the relationship between the total number of genes and the evolution of brain organization. The worm and the fruitfly genomes are similar in size, with respectively 18,424 and 13,601 different genes. But there are only 302 neurons in the nervous system of the worm and about 250,000 in that of the fly. The number of genes of the flowering plant *Arabidopsis* is even larger (about 25,706), even though it does not have a nervous system. Even more surprising, the number of genes from bony fish to laboratory mouse up to humans is roughly constant, around 30,000.

Moreover, from *Caenorhabditis* to humans, the total number of DNA bases in the genome increases about 30 times from 100 to 2,910 Kb, while the total number of genes increases only three to four times. As mentioned, the coding sequences (referred to as exons) span about only 1 percent of the total genome sequence. Most of the DNA is made up of repetitive or intervening sequences. The size of these noncoding sequences increases with evolution as noted, for instance, between mouse and man genomes (Mouse Genomic Sequencing Consortium, 2002). They may represent traces of the paleontological history but also may serve as active agents in the changes of shape of the genome (see IHGSC, 2001).

In conclusion, if from simpler organisms to vertebrates a simple relationship exists between the number of genes and the complexity of the organisms, it is lost in vertebrates, where the estimated total number of coding sequences remains constant while the complexity of brain organization dramatically increases in the course of evolution. How to solve this nonlinearity problem?

GENES AND THE MORPHOGENESIS OF THE BRAIN: FROM MONKEY TO HUMAN

The molecular genetics of the early stages of embryonic development in *Drosophila, Xenopus,* chick, mouse, and humans offer one major perspective to deal with both paradoxes of gene parsimony and of nonlinear evolution of gene-brain complexity.

First, in *Drosophila* a variety of genes have been identified that control the Cartesian coordinates of the embryo, the segmentation of the body and the identity of the segments (Driever and Nüsslein-Volhard, 1988). These developmental genes are, for example, the homeotic Hox genes whose mutation in the fly, for instance, causes the transformation of antennae into legs and thus controls the identity of a body segment. Closely related genes are found as well in vertebrate embryos. For instance, there are 160 homeobox domains in humans, 100 in the fly, 82 in the worm, but only 6 in yeast (Venter et al., 2001). Their "nonaccidental resemblance" suggests a common evolutionary origin of the ontogenetic pattern (Arendt and Nübler-Jung, 1999). This is true for the body plan but also for the brain coordinates.

In *Drosophila,* the Cartesian (head-tail, dorso-ventral, right-left) coordinates of the embryo are established very early, before the eggs are laid down. On the other hand, in mammals, they form rather late following the first cleavages of the fertilized egg. In mice, this occurs approximately 6.0 days after fertilization (Beddington & Robertson, 1999). At this stage, the head of the embryonic mouse is labeled by a small set of embryonic cells called the visceral endoderm.

Interestingly these pioneering cells express developmental genes (such as *OtX_2*) which are also found in *Drosophila* embryo (such as *Orthodenticle).* In addition, the mutation of such genes, for example *OtX_2* in the mouse, gives rise to an embryo that fails to develop a normal anteroposterior axis to the extent that the head itself does not form. In other words, *OtX_2* and homologs are necessary to make a head both in the fly and in the mouse. Thus despite different timing relationships, the patterns of genes expressed in the course of early development of the body look strikingly similar both in insects and vertebrates (Arendt and Nübler-Jung, 1997, 1999).

The genesis of the vertebrate brain in the course of evolution did not require an entire rebuilding from invertebrates. Gene duplications, possibly two successive duplications of the whole genome, may have sufficed. As was first articulated by Geoffroy Saint-Hilaire with his "unité de plan," a considerable number of structural determinants are saved from our invertebrate ancestors.

An important point to emphasize is that in the course of embryonic and postnatal development, these developmental genes become expressed according to well-defined spatiotemporal patterns, in a hierarchical and parallel manner with cross-regulatory interactions and reutilizations. Such a view of morphogenesis, as a developing network of gene interactions may account, at least in part, for the parsimony paradox. An enormous diversity, indeed, may arise from such combinatorial expression of a limited number of genes.

Also, in vertebrates, a striking asymmetry of the body organs (such as the liver and heart) as well as of the brain hemispheres (in mammals and particularly in humans) does exist. Single genes such as *lefty* in the mouse, or *situs inversus* in humans, together with their protein product (a protein molecule referred to as dynein, known to be involved in flagellar movement) have also been shown to determine the left-right asymmetry of the body plan (Beddington & Robertson, 1999). In all these instances, a few discrete genetic events dramatically change the overall pattern of developmental genes' expression, which directs the three-dimensional plan of the body and thus of the brain. The phenomenon of gene conservation is balanced by the nonlinear expression of a few of them.

In a general manner, at critical stages of embryonic development, "symmetry breakings" take place as manifested, for example, by the development of anteroposterior and dorsoventral polarities, of sharp boundaries between territories, and/or patterns of stripes. Allan Turing proposed in 1952 a theory that accounts for such features. On formal grounds, such defined and reproducible patterns might be generated from a set of chemical substances, or morphogens, which cross-react and diffuse throughout the organism. For instance, gradients of diffusible morphogens are thought to contribute to the unfolding of developmental gene expression, resulting in anteroposterior polarity (Meinhardt & Gierer, 1974). The main actors (but not only ones) are the products of the developmental genes: regulatory proteins, the already mentioned *transcription factors* that control gene transcription at the level of the core RNA polymerase II transcription complex (see Mannervik, Nibu, Zhang, & Leving, 1999). They bind to DNA elements (enhancers or silencers) that lock or unlock the transcription of adjacent structural genes and are themselves often conserved across species. Interplay between morphogens and transcription factors (coactivators and/or corepressors) build up an *intracellular network of gene regulation,* together with membrane receptors and the relevant second messengers. Such sets of molecules may contribute to the "reading" of a gradient of morphogen by an autocatalytic nonlinear switch in both a non-cellularized (Kerszberg & Changeux, 1994) or a cellularized embryo (Kerszberg, 1996). Such reading may even require a rather particular kind of *molecular interconnections* at the level of the transcription factors: for instance, the assembly of molecular partners into allosteric *hetero-oligomers* between one morphogen molecule from the gradient and a transcriptional coregulator now coded by a gene expressed in the embryonic nuclei. *Nonlinear* relationships between transcription factor concentration and morphogenesis may thus emerge from these combinatorial relationships, which include *autocatalytic switches.*

Differential combinations of transcription factors at the level of the promoters would give rise to sharp boundaries of gene expression at definite positions on the morphogen gradient (Kerszberg & Changeux, 1994). Computer simulations reveal that in addition to the formation of sharp and stable boundaries, the model accounts for the formation of patterns of bands distributed at well-defined positions along the gradient (see also Smolen, Baxter, & Byrne, 2000). The suggested combinatorial and nonlinear networks of transcription factors (Kerszberg & Changeux, 1994) have, since then, been experimentally documented in several examples of *Drosophila* development (Mannervik et al., 1999; Carmena et al., 1998; Halfon et al., 2000) among others (Davidson et al., 2002). Among other predictions, the model accounts for the displacement of the band pattern experimentally observed in *Drosophila* embryo as a function of transcription factor levels (Hoch, Seifert, & Jäckle, 1991). It illustrates how the activation of a given gene may depend on the "context" of genes expressed through the network of transcription factors which link them together.

Along these lines, one may note that what makes the difference between yeast and the nematode as a "minimal" multicellular organism are genes coding for transcription factors (e.g., 270 nuclear hormone receptors), protein-protein interaction domains (e.g., 156 POZ domains), or signal transduction domains (e.g., 11 phosphotyrosine binding domain) (Chervitz et al., 1998). In humans, an even more dramatic expansion of transcription factor genes takes place. For instance, the number of transcription factor domains referred to as CH2H2 zinc fingers increases from 771 in the fly to the huge number of 4,500 in humans. Consistent with our hypothesis of *transcription factor networks* formation, many such factors contain domains that are involved in the oligomerization of transcription factors. Some of them, like KRAB or SCAN, are even absent from the fly or worm genomes. Others result from the reshuffling or "accretion" of already existing invertebrate domains. These gene domains are expected to increase the combinatorial partnering of the transcription factors (Venter et al., 2001). Many of these new genes that I referred to in the past as "communication genes" (Changeux, 1983) are dedicated to the formation of patterns of gene interactions. Should we then say that what make us human lies in transcription factor genes and in the promoters to which they combine?

Many developmental genes are expressed in the nervous system. They are parts of a still largely uncharacterized population of genes concerned with brain morphogenesis from monkey or humans. The concept mentioned for embryonic development may apply as well to brain morphogenesis, in particular to its very early stage of formation referred to as neurulation (see Kerszberg & Changeux, 1998). Interestingly, the process of neurulation differs strikingly in invertebrates and in vertebrates. In the first case, the neuroblasts delaminate from the neural ectoderm to progressively form the solid ganglion chain of the adult (which may reach up to 520 million neurons in *Octopus*). In the second, the neural plate invaginates *en bloc* to form a hollow neural tube, which, as such, may facilitate a dramatic growth of the central nervous system through a surface expansion as is observed all the way from cyclostomes to mammals, primates, and humans. Such transitions may not require a large number of molecular changes at the gene transcription level. These genetic events may, for instance, affect transcription factor switches, which themselves regulate cell motion (Kerszberg & Changeux, 1998) and cell adhesion (see Edelman, 1988). As a consequence, either the whole neural tissue (neural plate) infolds into a tube (vertebrates), or individual neuroblasts delaminate yielding a solid nervous system (invertebrates). This illustrates in still highly hypothetical terms how a few gene changes may contribute to the critical transition between the invertebrate and vertebrate nervous system. A similar reasoning may apply to the shift of the position of the nervous system from ventral to dorsal which distinguishes invertebrates from vertebrates.

Even less understood is the expansion of the cerebral cortex that took place in the course of vertebrate brain evolution, in particular from monkey to man (see Mountcastle, 1998; Changeux, 1983). The number of neurons per cortical column is rather uniform throughout the vertebrates (Rockell, Hiorns, & Powell, 1980). Thus the surface area of the cortex, i.e., the number of columns, appears as the primary target of the evolutionary changes (Rakic, 1988). The gestation lasts 21 days in the rat, 165 in the macaque, and 280 in humans, and the rapid phase of synaptogenesis (which starts two months before birth in macaque and four to five months after birth in man) lasts 136 days in macaque and 470 days in humans. One may further speculate that the fast expansion of the frontal lobe and parietotemporal areas, which contributed to the evolutionary origins of *Homo sapiens'* brain, resulted from the exceptionally prolonged action of some still unidentified developmental genes (Changeux, 1983), the genomic evolution underlying this process engaging a rather small set of genes or regulatory sequences. An important consequence of the increased surface of the cerebral cortex is the increased possibility of establishing connections between cortical neurons.

As discussed (Changeux, 1983), this translates into an increase in the mean number of connections per neurons, with a consequent burgeoning of the dendritic and axonal trees. For instance, the average number of synapses per neuron increases from 2,000 to 5,600 in the monkey up to 6,800 to 10,000 in humans (Bourgeois, 2003; see Ramon y Cajal, 1910-1911). This is especially true for the prefrontal association cortex, which differentially expands in the human lineage. Indeed, according to Elston (2003), human prefrontal pyramidal cells, which have, on the average, 15,100 spines in their basal arbors, show 72 percent more spines than those in macaque, 3.8 times more than in marmoset, and 7.5 times more than those in

Aotus. Also, the differential increase in complexity concerns more the prefrontal association cortex than the primary cortical areas: for instance, the prefrontal pyramidal neurons are on the average 16 times more spinous than those in V1 (Elston, 2000). Moreover, von Economo's pioneering observations on the architectonics of cortical areas (1929) underlined that layer II-III pyramidal neurons are more abundant in prefrontal and parietotemporal association areas than in the rest of the cerebral cortex. Since the layer II-III neurons are known to possess long-range cortico-cortical and callosal axons, the differential expansion of the prefrontal cortex is accompanied by an increase of long range connectivity and in a general manner of the white matter (Olesen, Nagy, Westerberg, & Klingberg, 2003). These convergent findings are thus consonant with the proposal that the long-range cortico-cortical connexions contribute to the neural architectures of a global "conscious workspace" (Dehaene, Kerszberg, & Changeux, 1998; Baars, 1988) in which performances increase from monkey to human (see Dehaene, Sergent, & Changeux, 2003), and which develops in the course of child development (Olesen et al., 2003).

Can we identify the particular genes that cause this neocortical surface expansion? Close comparison of the gene repertoires between mouse and human and among mammals in general reveals a high degree of conservation (Mouse Genomic Sequencing Consortium, 2002). Even more surprisingly, some particular gene families—like the olfactory receptor gene family—show a decrease of functional genes in humans compared to chimpanzees. Humans have more than 1,000 olfactory receptor genes, but only approximately 40 percent have an intact coding region and are therefore functional. In contrast, the fraction of intact olfactory genes in the genomes of the great apes is significantly greater (68-72 percent) (Gilad, Bustamante, Lancet, & Pääbo, 2003). We have to consider the unexpected possibility that from early mammals to humans, the number of genes may have decreased!

Overall, looking for genetic changes in the coding sequence of proteins from chimpanzees to man, according to Carroll (2003) the most extensive comparison indicates an average substitution level of about 1.2 percent in single-copy DNA. In addition, gene duplications, chromosomic reorganizations, insertions, and deletions might have to be considered. The total number of amino acid replacements that contributed to protein evolution in the human lineage is estimated to be of the order of 200,000. Yet, one does not know which fraction caused functional changes. The number of so- called adaptative substitutions would then be even smaller: between 10,000 adapatative substitutions in human genes and their regulatory regions (Carroll, 2003). Are all of them unique to humans and their cognitive functions? Which ones are candidates for being causally implicated in past human evolution? In a recent provocative report mutations of the gene FOX P2 (forkhead box P2) were found associated with speech and language disorder (Lai et al., 2001; Enard et al., 2002; Lai et al., 2003). They cause severe articulation difficulties accompanied by linguistic and grammatical impairment. Is this the long-expected "gene for language"? In fact, this is not the case. It is found in other species, and the human FOX P2 differs from the gorilla and chimpanzee sequence at just two residues and from the mouse sequence at four residues (Enard et al., 2002). FOX P2 is a transcription factor, and one may anticipate that, as mentioned above, it contributes to complex regulatory networks of gene interactions. Its inactivation may, among other factors, cause a disorganization of the network of transcription factors (see Kerszberg & Changeux, 1994, 1998).

Even if the number of genetic changes that took place in the human lineage looks rather small, even if it ultimately boils down to a few thousand, the relationships between these genes and the evolution of neural architecture from monkey to humans cannot be thought of as straightforward. Multiple nested networks of interactions are expected to develop among gene products, in space and time, in the course of human body and brain formation. As a consequence, single-gene actions are expected to be altogether pleiotropic: they affect multiple aspects of the functional organization of the brain (and even of the body). Furthermore, they are context-dependent, and are therefore anticipated to differ in different tissues or organs and at different stages of development. To talk about "language genes" or "genes of intelligence" no longer seems appropriate. But to decipher the regulatory network of transcription factors that lead to the differential expansion of the neocortex and, more specifically, of the pre- frontal cortex begins to appear as a plausible challenge for the next decade.

THE ACTIVITY-DEPENDENT EPIGENESIS OF NEURONAL NETWORKS AND THE ORIGIN OF CULTURE

In the preceding chapters, the paradox of nonlinear evolution between genome and brain complexity has been discussed, exclusively, on the basis of gene interactions. A straightforward prediction of this scheme is that genetically identical, individuals—identical twins or clones—should be neurally identical. In reality, this is not the case. An important variability of the organization of the nervous system exists between genetically identical individuals. For instance, in vivo measurements of the *planum temporale* by magnetic resonance and the results of behavioral tasks collected with monozygotic co-twins discordant for manuality yielded convergent results. The righthanders showed leftward hemispheric asymmetry, whereas the left-handers lacked asymmetry (Steinmetz et al., 1995; Eckert et al., 2002). Early "epigenetic" events take place during embryogenesis and contribute to a significant variability in the development of the anatomo-functional laterality of the cerebral hemispheres. At the neuronal level, a phenotopic variance has been identified in cloned parthenogenetic individuals from the water flea *Daphnia magna* and the fish *Poecilia formosa* (see Levinthal, Macagno, & Levinthal, 1976). At the electron microscope level, the details of the axonal or dendritic branching of an identifiable neuron differ among genetically identical individuals, and the within-individual variability between left and right arborization is smaller than that found from between individuals. In other words, "cloned" individuals are not neurally identical, even in species with a fixed number of identifiable nerve cells. In a mammal such as the mouse (the situation might be even more extreme in humans) the number of cells is much larger, and there are no longer identifiable single cells. Despite common principles of neural architecture delimited by a species-specific genetic envelope, the individual variability of the fine anatomy observed between individuals from genetically homogeneous lineages increases dramatically.

One plausible solution (among many others) is that the state of activity of the developing nervous system contributes to the organization of the adult network by trimming up synapse formation at sensitive periods of development. The model suggested (Changeux, Courrège, & Danchin, 1973; Changeux, 1983; see also Elliott & Shadbolt, 1998) posits that during synapse formation the genetic envelope controls—in addition to the division, migration, and differentiation of cell categories—the behavior of the growth cone, the outgrowth and formation of widespread connections, the recognition of the target cells, and the onset of spontaneous activity; it also determines the structure and function of the molecules that enter into the architecture of the synapse, the rules governing their assembly, and their evolution in relation to the activity of the network. Yet, at sensitive periods of circuit development, the phenotypic variability of nerve cell distribution and position, as well as the exuberant spreading and the multiple figures of the transiently formed connections originating from the comings and goings of growth cone behavior, introduce a *maximal diversity* that is then reduced by the "active" selective stabilization of some of the labile contacts and the elimination (or retraction) of the others at subsequent stages of development. The crucial hypothesis of the synapse elimination model (Changeux, Courrège, & Danchin, 1973) is that the evolution of the connective state of each synaptic contact is governed globally and within a given time window by the overall signals received from the cell on which it terminated. In other words, the activity of the postsynaptic cell regulates the stability of the synapse in a *retrograde manner.*

The contribution of neural activity (evoked and/or spontaneous) in the formation of cortical circuits has been documented since the classic experiments of Wiesel and Hubel (1963), which demonstrated the important role of visual experience in fixing the organization of ocular dominance columns (see Katz & Shatz, 1996). The exuberant sprouting and proliferation of axonal branches, accompanied by limited though critical elimination of collaterals, has been visualized at different locations along the visual pathway (retinogeniculate, thalamocortical, and pyramidal cell arbors) (see Sretavan & Shatz, 1986; Katz & Shatz, 1996) at sensitive periods of development. It is clear that the state of activity of the developing cortical circuit controls synaptic evolution by gestation and continues actively during the first two postnatal years. In humans, the *mean synaptic density* reaches a maximum near postnatal age 3 months in the auditory cortex but only *after* age 15 months in the late maturing middle frontal gyrus (Huttenlocher & Dabholkar, 1997). A phase of *net decrease* in mean synaptic density occurs subsequently in childhood, where it ends earlier in auditory cortex (until 12 years) than in prefrontal cortex (until mid-adolescence). Most likely, this global evolution of the mean synaptic density represents the summation of "nested" waves of synapse outgrowth and elimination, the later process becoming dominant and thus visible in global measures of net synaptic density late in development (Huttenlocher & Dabholkar, 1997; Bourgeois 2001, 2003).

The example of the neuromuscular junction is particularly simple since only a single synaptic contact persists in the adult, while in the newborn rat each fiber receives four or five *active* motor axon terminals. As the rat begins to walk, the number of functional terminals progressively decreases until for the adult only one is left. The state of activity of the innervated muscle controls this elimination (e.g., Benoit and Changeux, 1975, 1978; see Sanes and Lichtman, 1999). Similar regressive phenomena have also been documented at the synaptic level in other systems such as the sympathetic ganglia (Purves & Lichtman, 1980) or the climbing fiber Purkinje cell synapse in the cerebelum (see Crépel, Mariani, & Delhaye-Bouchaud, 1976; Changeux & Mikoshiba, 1978; Kano et al., 1997). For the latter, a mutation which inactivates a specific neurotransmitter receptor (the type 1 metabotropic glutamate receptor mGluR1) delays the regression of supernumerary climbing fiber innervation (Kano et al., 1997).

Among the various consequences of our modeling approach, a simple one has been to look for the molecular mechanisms of synaptic outgrowth, stabilization, and elimination and their regulation by neural activity. At the presynaptic level, neurotrophins like NGF, BDNF, NT4 and several others (see Levi-Montalcini, 1987; Barde, 1990) have become plausible candidates for retrograde signals in the activity-dependent-synaptic outgrowth and selection (Thoenen, 1995; Katz & Shatz, 1996). For instance, in vivo intracortical infusion of diverse trophins prevents the shift of ocular dominance in favor of the nondeprived eye (Maffei et al., 1992; Carmignoto et al., 1993) on the formation of ocular dominance columns (Cabelli, Hohn, & Schatz, 1995). Also, rapid and opposite effects of BDNF and NGF have been demonstrated on the whisker barrel representation of the rat somatosensory cortex. Moreover, neurotrophins modulate synaptic strength within minutes in vitro at cultured neuromuscular synapses (Lohof, Ip, & Poo, 1993) and GDNF overexpression in the mouse delays the elimination of supernumerary motor axons and causes hyperinnervation of neuromuscular junctions (Nguyen et al., 1998). Formal models of synapse-selective stabilization based on competition for limited stocks of trophic factors have thus been developed (Gouzé, Lasry, & Changeux, 1983; Kerszberg, Denaene, & Changeux, 1992; Elliott & Shadbolt, 1998), giving even more plausibility to the theory. A molecular genetic approach of activity-dependent synapse selection thus becomes plausible (Schaeffer, Duclert, Huchet-Dymanus, & Changeux, 1998; Schaeffer, de Kerchove d'Exaerde, & Changeux, 2001; de Kerchove d'Exaerde et al., 2002), raising the issue of the relative contribution of these epigenetic processes to the complexity of the brain (see Changeux, 1983).

In mammals a striking relationship exists between the length of gestation and the complexity of the brain: 21 days in the rat, 65 in the cat, 165 in the macaque, and 280 in man. Synaptogenesis in the neocortex of the newborn has been subdivided by Bourgeois (2001) in four phases. Phases 1 and 2 correspond to low density early phases. Phase 3 is rapid and accounts for the overall maximal density of synapses in the cortex. It takes place postnatally in the rat and cat (2–19 days respectively *after* birth) but starts before birth in the monkey and human (2 months and 4–5 months respectively before delivery). But, most significant for us is the increased duration of this phase 3: 14 days in the rat, 30 in the cat, 136 in the monkey, and 470 in the human, thus an increase by at least thirtyfold from rat to human. Bourgeois (2001) has noted that the early phases of synaptogenesis 1 and 2 are independent of the environment, while the subsequent phases 3 and 4 require the interaction with the outside world to develop normally. There is thus a postnatal extensive prolongation of the experience-sensitive phases of synapse formation in the infant brain, which has major consequences. The brain of the newborn human is exposed to the physical, social, and cultural environment for extended period of times compared to the monkey. As a consequence, it becomes accessible to "cultural imprints" such as those involved in spoken and written language learning. A culture may thus develop. As discussed (Changeux, 1983; Carroll 2003), a few mutations, for example on regulatory genes controlling hormonal release, may have caused this increased duration and thus simply made possible the onset of human culture.

Brain epigenetic capacities to store stable representations of the outside world give human beings the opportunity to create an artificial world of cultural objects at the social level. In other words, the origin of culture and of its transmission from generation to generation lies in the considerable increase of synapse numbers and multiple nested processes of activity-dependent synapse selection that take place postnatally in the human brain. This epigenetic evolution also has another consequence: it permits the diversification of the cultures that human beings have developed throughout their recent history. In other words, the postnatal epigenetic evolution of brain connectivity opens the way to cultural evolution.

REFERENCES

Aboitiz, F., & Garcia, R. (1997). The anatomy of language revisited. *Biological Research, 30,* 171-183.

Adams, M. D., Celniker, S. E., Holt, R. A., Evans, C. A., Gocayne, J. D., Amanatides, P. G., et al. (2000). The genome sequence of Drosophila melanogaster. *Science, 287,* 2185-2195.

Aparicio, S., Chapman, J., Stupka, E., Putnam, N., Chia, J. M., Dehal, P., et al. (2002). Whole-genome shotgun assembly and analysis of the genome of Fugurubripes. *Science, 297,* 1301-1310.

Arendt, D., & Nübler-Jung, K. (1997). Dorsal or ventral: similarities in fate maps and gastrulation patterns in annelids, arthropods and chordates. *Mechanical Development, 61,* 7-21.

Arendt, D., & Nübler-Jung, K. (1999). Rearranging gastrulation in the name of yolk: Evolution of gastrulation in yolk-rich amniote eggs. *Mechanical Development, 81,* 3-22.

Arabidopsis Genome Initiative. (2000). Analysis of the genome sequence of the flowering plant Arabidopsis thaliana. *Nature, 408,* 796-815.

Baars, B. (1988). *A Cognitive Theory of Consciousness.* New York: Cambridge University Press.

Barde, Y. A. (1990). The nerve growth factor family. *Progress in Growth Factor Research, 2,* 237-248.

Beddington, R. S., & Robertson, E. J. (1999). Axis development and early asymmetry in mammals. *Cell, 96,* 195-209.

Benoit, P., & Changeux, J. P. (1975). Consequences of tenotomy on the evolution of multiin-nervation in developing rat soleus muscle. *Brain Research, 99,* 354-358.

Benoit, P., & Changeux, J. P. (1978). Consequences of blocking the nerve with a local anaesthetic on the evolution of multiinnervation at the regenerating neuromuscular junction of the rat. *Brain Research, 149,* 89-96.

Bourgeois, J. P. (2001). Synaptogenesis in the neocortex of the newborn: the ultimate frontier for individuation? In H. Lagercrantz, M. Hanson, P. Evrard, & C. Rodeck (Eds.), *The Newborn Brain: Neuroscience and Clinical Applications* (pp. 91-113). Cambridge: Cambridge University Press.

Bourgeois, J. P. (2003). Le développement de la connectivité cérébrale: étape ultime de l'individuation? In J. P. Changeux (Ed.), *Gènes et culture* (93-116). Paris: Odile Jacob.

Brodmann, K. (1912). Neue Ergebnisse über die vergleichende histologische Lokalisation der grosshirnrinde Mitbesonderer Berücksichtigung des Stirnhirns. *Anatomische Anzeiger, 41,* 157-216.

C. elegans Sequencing Consortium. (1998). Genome sequence of the nematode C. elegans: A platform for investigating biology. *Science, 282,* 2012-2018. Review. Erratum in. *Science,* 283 (5398), 35; *Science 283* (5410), 2103; *Science 285* (5433), 1493.

Cabelli, R. J., Hohn, A., & Shatz, C. J. (1995). Inhibition of ocular dominance column formation by infusion of NT-4/5 or BDNF. *Science, 267,* 1662-1666.

Cantalupo, C., & Hopkins, W. D. (2001). Asymmetric Broca's area in great apes. *Nature, 414,* 505.

Carmena, A., Gisselbrecht, S., Harrison, J., Jimenez, F., & Michelson, A. M. (1998). Cell fates in the Drosophila embryonic mesoderm. *Gene Development, 12,* 3910-3922.

Carmignoto, G., Canella, R., Candeo, P., Comelli, M. C., & Maffei, L. (1993). Effects of nerve growth factor on neuronal plasticity of the kitten visual cortex. *Journal of Physiology, 464,* 343-360.

Carroll, S. B. (2003). Genetics and the making of Homo sapiens. *Nature, 422,* 849-857.

Changeux, J. P. (1983). *L'Homme neuronal.* Paris: Fayard.

Changeux, J. P., Courrege, P., & Danchin, A. (1973). A theory of the epigenesis of neuronal networks by selective stabilization of synapses. *Proceedings of the National Academy of Sciences, 70,* 2974-2978.

Changeux, J. P., & Mikoshiba, K. (1978). Genetic and "epigenetic" factors regulating synapse formation in vertebrate cerebellum and neuromuscular junction. *Progress in Brain Research, 48,* 43-66.

Chervitz, S. A., Aravind, L., Sherlock, G., Ball, C. A., Koonin, E. V., Dwight, S. S., et al. (1998). Comparison of the complete protein sets of worm and yeast: orthology and divergence. *Science, 282,* 2022-2028.

Crepel, F., Mariani, J., & Delhaye-Bouchaud, N. (1976). Evidence for a multiple innervation of Purkinje cells by climbing fibers in the immature rat cerebellum. *Journal of Neurobiology, 7,* 567-578.

Davidson, E. H., Rast, J. P., Oliveri, P., Ransick, A., Calestani, C., Yuh, C. H., et al. (2002). A genomic regulatory network for development. *Science, 295,* 1669-1678.

de Kerchove d'Exaerde, A., Cartaud, J., Ravel-Chapuis, A., Seroz, T., Pasteau, F., Angus, L. M., et al. (2002). Expression of mutant Ets protein at the neuromuscular synapse causes alterations in morphology and gene expression. EMBO Rep., *3,* 1075-1081.

Deacon, T. (1997). *The Symbolic Species.* New York: Norton.

Dehaene, S., & Changeux, J. P. (2000). Reward-dependent learning in neuronal networks for planning and decision making. *Progress in Brain Research, 126,* 217-229.

Dehaene, S., Kerszberg, M., & Changeux, J. P. (1998). A neuronal model of a global workspace in effortful cognitive tasks. *Proceedings of the National Academy of Sciences, 95,* 14529-14534.

Dehaene, S., Sergent, C., & Changeux, J. P. (2003). A neuronal network model linking subjective reports and objective physiological data during conscious perception. *Proceedings of the National Academy of Sciences,* 100(14), 8520-8525.

Driever, W., & Nusslein-Volhard, C. (1988). The bicoid protein determines position in the Drosophila embryo in a concentration-dependent manner. *Cell, 54,* 95-104.

Eckert, M. A., Leonard, C. M., Molloy, E. A., Blumenthal, J., Zijdenbos, A., & Giedd, J. N. (2002). The epigenesis of planum temporale asymmetry in twins. *Cerebral Cortex, 12,* 749-755.

Edelman, G. M. (1988). Morphoregulatory molecules. *Biochemistry,* 27(10), 3533-3543.

Elliott, T., & Shadbolt, N. R. (1998). Competition for neurotrophic factors: mathematical analysis. *Neural Computation, 10,* 1939-1981.

Elston, G. N. (2000). Pyramidal cells of the frontal lobe: All the more spinous to think with. *Journal of Neuroscience, 20,* RC95.

Elston, G. N. (2003). Cortex, cognition and the cell: New insights into the pyramidal neuron and prefrontal function. *Cerebral Cortex, 13,* 1124-1138.

Enard, W., Przeworski, M., Fisher, S. E., Lai, C. S., Wiebe, V., Kitano, T., Monaco, A. P., & Pääbo, S. (2002). Molecular evolution of FOXP2, a gene involved in speech and language. *Nature, 418,* 869-872.

Gannon, P. J., Holloway, R. L., Broadfield, D. C., & Braun, A. R. (1998). Asymmetry of chimpanzee planum temporale: Humanlike pattern of Wernicke's brain language area homolog. *Science, 279,* 220-222.

Gilad, Y., Bustamante, C. D., Lancet, D., & Pääbo, S. (2003). Natural selection on the olfactory receptor gene family in humans and chimpanzees. *American Journal of Human Genetics, 73,* 489-501.

Goffeau, A., Barrell, B. G., Bussey, H., Davis, R. W., Dujon, B., Feldmann, H., et al. (1996). Life with 6,000 genes. *Science, 274*(5287), 546, 563-567.

Gouze, J. L., Lasry, J. M., & Changeux, J. P. (1983). Selective stabilization of muscle innervation during development: A mathematical model. *Biological Cybernetics, 46,* 207-215.

Halfon, M. S., Carmena, A., Gisselbrecht, S., Sackerson, C. M., Jimenez, F., Baylies, M. K., & Michelson, A. M. (2000). Ras pathway specificity is determined by the integration of multiple signal-activated and tissue-restricted transcription factors. *Cell, 103,* 63-74.

Hoch, M., Seifert, E., & Jackle, H. (1991). Gene expression mediated by cis-acting sequences of the Kruppel gene in response to the Drosophila morphogens bicoid and hunchback. *EMBO Journal, 10,* 2267-2278.

Huttenlocher, P. R., & Dabholkar, A. S. (1997). Regional differences in synaptogenesis in human cerebral cortex. *Journal of Computational Neurology, 387,* 167-178.

International Human Genome Sequencing Consortium (2001). Initial sequencing and analysis of the human genome. *Nature, 409,* 860-921. Erratum in Nature *412* (6846), 565; *Nature 411* (6838), 720.

Kano, M., Hashimoto, K., Kurihara, H., Watanabe, M., Inoue, Y., Aiba, A., & Tonegawa, S. (1997). Persistent multiple climbing fiber innervation of cerebellar Purkinje cells in mice lacking mGluR1. *Neuron, 18,* 71-79.

Katz, L. C., & Shatz, C. J. (1996). Synaptic activity and the construction of cortical circuits. *Science, 274,* 1133-1138.

Kerszberg, M. (1996). Accurate reading of morphogen concentrations by nuclear receptors: A formal model of complex transduction pathways. *Journal of Theoretical Biology, 183,* 95104.

Kerszberg, M., & Changeux, J. P. (1994a). A model for reading morphogenetic gradients: auto- catalysis and competition at the gene level. *Proceedings of the National Academy of Sciences, 91* , 5823-5827.

Kerszberg, M., & Changeux, J. P. (1994b). Partners make patterns in morphogenesis. *Current Biology, 4,* 1046-1047.

Kerszberg, M., & Changeux, J. P. (1998). A simple molecular model of neurulation. *Bioessays, 20,* 758-770.

Kerszberg, M., Dehaene S., & Changeux J. P. (1992). Stabilization of complex input-outpout functions in neural clusters formed by synapse selection. *Neural Networks, 5,* 403-413.

Lai, C., Fisher, S. E., Hurst, J. A., Vargha-Khadem, F., & Monaco, A. P. (2001). A forkhead-domain gene is mutated in a severe speech and language disorder. *Nature, 413,* 519-523.

Lai, C. S., Gerrelli, D., Monaco, A. P., Fisher, S. E., & Copp, A. J. (2003). FOXP2 expression during brain development coincides with adult sites of pathology in a severe speech and language disorder. *Brain, 126,* 2455-2462.

Levi-Montalcini, R. (1987). The nerve growth factor 35 years later. *Science, 237,* 1154-1162.

Levinthal, F., Macagno, E., & Levinthal, C. (1976). Anatomy and development of identified cells in isogenic organisms. *Cold Spring Harbor Symposia in Quantitative Biology, 40,* 321-331.

Lohof, A. M., Ip, N. Y., & Poo, M. M. (1993). Potentiation of developing neuromuscular synapses by the neurotrophins NT-3 and BDNF. *Nature, 363,* 350-353.

Maffei, L., Berardi, N., Domenici, L., Parisi, V., & Pizzorusso, T. (1992). Nerve growth factor (NGF) prevents the shift in ocular dominance distribution of visual cortical neurons in monocularly deprived rats. *Journal of Neuroscience, 12,* 4651-4662.

Mannervik, M., & Levine, M. (1999). The Rpd3 histone deacetylase is required for segmentation of the Drosophila embryo. *Proceedings of the National Academy of Sciences, 96,* 6797-6801.

Mannervik, M., Nibu, Y., Zhang, H., & Levine, M. (1999). Transcriptional coregulators in development. *Science, 284,* 606-609.

Meinhardt, H., & Gierer, A. (1974). Applications of a theory of biological pattern formation based on lateral inhibition. *Journal of Cell Science, 15,* 321-346.

Mountcastle, V. (1998). *Perceptual neuroscience: The cerebral cortex.* Cambridge, MA: Harvard University Press.

Mouse Genomic Sequencing Consortium. (2002). Initial sequencing and comparative analysis of the mouse genome. *Nature, 420,* 520-562.

Nguyen, Q. T., Parsadanian, A. S., Snider, W. D., & Lichtman, J. W. (1998). Hyperinnervation of neuromuscular junctions caused by GDNF overexpression in muscle. *Science, 279,* 1725-1729.

O'Donovan, C., Apweiler, R., & Bairoch, A. (2001). The human proteomics initiative (HPI). *Trends in Biotechnology, 19,* 178-181.

Olesen, P. J., Nagy, Z., Westerberg, H., & Klingberg, T. (2003). Combined analysis of DTI and fMRI data reveals a joint maturation of white and grey matter in a fronto-parietal network. *Brain Research/Cognitive Brain Research, 18,* 48-57.

Purves, D., & Lichtman, J. W. (1980). Elimination of synapses in the developing nervous system. *Science, 210,* 153-157.

Rakic, P. (1988). Specification of cerebral cortical areas. *Science, 241,* 170-176.

Rauschecker, J. P., Tian, B., & Hauser, M. (1995). Processing of complex sounds in the macaque nonprimary auditory cortex. *Science, 268,* 111-114.

Rockel, A. J., Hiorns, R. W., & Powell, T. P. (1980). The basic uniformity in structure of the neocortex. *Brain, 103,* 221-244.

Rubin, G. M., Yandell, M. D., Wortman, J. R., Gabor Miklos, G. L., Nelson, C. R., Hariharan, I. K., et al. (2000). Comparative genomics of the eukaryotes. *Science, 287,* 2204-2215.

Sanes, J. R., & Lichtman, J. W. (1999). Development of the vertebrate neuromuscular junction. *Annual Reviews of Neuroscience, 22,* 389-442.

Schaeffer, L., de Kerchove d'Exaerde, A., & Changeux, J. P. (2001). Targeting transcription to the neuromuscular synapse. *Neuron, 31,* 15-22.

Schaeffer, L., Duclert, N., Huchet-Dymanus, M., & Changeux, J. P. (1998). Implication of a mul- tisubunit Ets-related transcription factor in synaptic expression of the nicotinic acetylcholine receptor. *EMBO J., 17,* 3078-3090.

Semendeferi, K., Armstrong, E., Schleicher, A., Zilles, K., & Van Hoesen, G. W. (2001). Prefrontal cortex in humans and apes: A comparative study of area 10. *American Journal of Physical Anthropology, 14,* 224-241.

Semendeferi, K., Damasio, H., Frank, R., & Van Hoesen, G. W. (1997). The evolution of the frontal lobes: A volumetric analysis based on three-dimensional reconstructions of magnetic resonance scans of human and ape brains. *Journal of Human Evolution, 32,* 375-388.

Semendeferi, K., Lu, A., Schenker, N., & Damasio, H. (2002). Humans and great apes share a large frontal cortex. *Nature Neuroscience, 5,* 272-276.

Smolen, P., Baxter, D. A., & Byrne, J. H. (2000). Mathematical modeling of gene networks. *Neuron, 26,* 567-580.

Snyder, M., & Gerstein, M. (2003). Genomics: Defining genes in the genomics era. *Science, 300,* 258-260.

Sretavan, D., & Shatz, C. J. (1986). Prenatal development of cat retinogeniculate axon arbors in the absence of binocular interactions. *Journal of Neuroscience, 6,* 990-1003.

Steinmetz, H., Herzog, A., Schlaug, G., Huang, Y., & Jancke, L. (1995). Brain (A) symmetry in monozygotic twins. *Cerebral Cortex, 5,* 296-300.

Thoenen, H. (1995). Neurotrophins and neuronal plasticity. *Science, 270,* 593-598.

Venter, J. C., et al. (2001). The sequence of the human genome. *Science, 291,* 1304-1351. Erratum in *Science 292* (5523), 1838.

Von Economo, C. (1929). *The Cytoarchitectonics of the Human Cerebral Cortex.* Oxford: Oxford University Press.

Wiesel, T. N., & Hubel, D. H. (1963). Effects of visual deprivation on morphology and physiology of cells in the cats lateral geniculate body. *Journal of Neurophysiology, 26,* 978-993.

ARE THE NEANDERTALS

Finally Becoming Human?

By Nasser Malit

This chapter deals with the history of the Neandertals of Europe and the interpretation of their morphology and behavior. The goal here is to highlight how current knowledge and analyses, including the evidence from morphology, genetics, and behavior, may help to address matters of misrepresentation of this human group since the early 1800s until recent times. Earlier views on these fossil humans were largely erroneous, most likely due to a lack of existing discipline or framework for interpreting the evidence. Also, sociocultural factors at the time—especially the European views of the others (non-Europeans)—and the idiosyncrasies of early researchers may have had some influence on paleoanthropology at its inception. These negative ideas about Neandertals permeated the popular culture and continue to do so to this day. The science of paleoanthropology was also influenced by these prior ideas, but there has been a trend toward more positive views.

The Neandertals represent a phase in human evolution that is critical to our understanding of human diversity today. The fossil record of this enigmatic human species was from 1829 in Engis Cave, Belgium, but it was not until the discovery of the remains in the Neander Valley, Germany, in 1856 that actual study of this group began. It is important to note that the history of the Neandertals synchronizes with the chronological development of the science of human origins (paleoanthropology) as we know it today, and serves as a good case of how human evolutionary studies have passed through several phases that have improved theory, methods, and research approaches. This chapter will highlight some of the key biological and behavioral characteristics of the Neandertals and show how the these attributes of morphology and behavior may lead to a significant revision of the manner in which these humans are viewed today. The hope is that these new depictions will seep into popular culture and the media. The correlation of morphology, genes, and the archaeological record left by the Neandertals will be utilized for the analyses in this chapter. The biological characteristics will drive the discussion of the life ways of the Neandertals and help to revise the picture by comparison of their traits with those of modern humans. Current reconstruction of the behavioral evidence also provides further support for the need to revise completely the earlier biased views of the Neandertals. It is only through these approaches that an objective image may replace the previously tainted and prejudiced negative Neandertal stereotype.

INTRODUCTION

Neandertals are an extinct human group that lived throughout much of Europe and western Asia approximately 300,000 to 30,000 years before present (ky BP). The terms Neandertal, Neanderthal, or Neanderthaloid are derived from the Neander Valley (German, *Thal*) near Düsseldorf in Germany, where the "first" fossil specimen representing this group was discovered. The German orthographers later dropped the letter "h" in their spelling. It is, however, still possible to find books that use the letter "h" in their spelling. This chapter will use Neandertal throughout. *[Editor's Note: Brian M. Fagan, author of numerous textbooks on world archaeology, utilizes the earlier spelling; many others do not.]*

In 1856, at Feldhofer Cave, Neander Valley near Düsseldorf, Germany, a quarry crew working for Wilhelm Beckershoff accidentally discovered skeletal fragments of a fossil human. The partial skeleton was composed of a fragment of a skull cap (calotte), a right shoulder blade (scapula), a right collar bone (clavicle), both right and left upper arm bones (humeri), three lower arm bones (two ulnas and one radius), five ribs, the left half of a hip bone (pelvis), and two thigh bones (femuri). Beckershoff , the owner of the quarry, believed the bones were those of a cave bear. He later sought the opinion of a local teacher and fossil collector, Johann K. Fuhlrott of Eberfied, whom he asked to look at the bones. Fuhlrott's initial comments on the fossils were that the skull probably belonged to a bear, but the postcranials were undisputedly those of a human. Fuhlrott later reconsidered his initial view and described the fossils as those of an unusual human. He further indicated that the cranium had "unusually heavy bones, large brow-ridges which are completely joined in the middle." He proposed an Ice Age date for the fossils and concluded the fossils belonged to a "primitive member of our race." These initial observations generated such controversy that even the editors of the Society's Proceedings reportedly commented that they "cannot share the reported opinions."

This was the era of creationism and belief in the "Great Chain of Being." Therefore, it was not possible to fathom that humans were part of nature and subject to evolutionary forces. Evolutionary theory was still in its inception. Reacting to opposition to his conclusions on the fossils, Johann Fuhlrott warned "those attempting at propaganda" to leave the final judgment about the existence of the fossil man to the future. As a teacher, Fulhrott did not command media or scientific influence needed to sway those in power. His stature in the scientific community could not budge the opinion of the scholarly community. Later, he sought the opinion of Hermann Schaaffhausen, an anatomist from the University at Bonn, , who had discussed the existence of fossil humans in a publication *On the Stability and Change in Species* (1853). Fulhrott and Schaaffhausen later described the specimens and jointly presented their findings in July 1857 to the Natural History Society of the Prussian Rhineland. In 1859, Schaaffhausen published the seminal *Human Remains from a Cliff Grotto in the Düssel Valley*. The work largely agreed with Fuhlrott's original descriptions. Schaaffhausen's description emphasized the uniqueness of the fossils, the massive bones, and notable muscular development, adding that the individual was perhaps of Diluvian Age. He believed the fossils represented a race that had inhabited northern Europe before the Germans and the Celts, and recognized the possible arduous life circumstances of these raw and wild early Europeans. As a proponent of organic evolution, Schaaffhausen still did not explore the evolutionary implications of a Neandertal ancestral grouping.

August Mayer, a colleague of Schaaffhausen at Bonn, made observations about the Feldhofer specimens that would influence the perception of the Neandertals, which lasts to this day. His ideas were endorsed by the powerful Rudolf Virchow, who was then a well-known and influential Prussian pathologist and anatomist at the University at Bonn. Mayer asserted that the remains belonged to a modern human afflicted with rickets and interpreted the badly healed fracture on the arm to have caused intense pain that led to a constant frown, hence the enlarged brow ridges of the forehead. August Mayer's observation of the thigh bones led him to reason that the curvature indicated the thighs were of a lifelong equestrian and attributed their unique thickness and shape as due to rickets. He concluded that the bones were of an unfortunate deserter from a Russian Cossack cavalry suffering from rickets. Other afflictions, too, made him have a permanent frown. Mayer's analysis was endorsed by Rudolf Virchow after he personally looked at the original specimens in 1872. Virchow's conclusion was disappointing for a pathologist who was an expert on the identification of rickets. A major symptom of the condition is thinning, not thickening of the bones. Science historians now believe that this unfortunate position taken by Rudolf Virchow was due to his political motivation, driven by his antievolution sentiments.

Even Charles Darwin, a natural historian and evolutionist, completely ignored Neandertals or human evolution in his *Origin of the Species* (1859) due to its sensitivity at the time, although he was aware of the Neandertals' existence. Darwin avoided these as touchy issues that he later dealt with in his book, *The Descent of Man* (1871). On the other hand, Charles Lyell, a geologist who introduced the theory of uniformitarianism and deep time in his work, *The Geological Evidence of the Antiquity of Man* (1863), stressed the great antiquity implied by fossil animals that accompanied the Engis Cave Neandertal child and the degree of fossilization of the Feldhofer remains. Being a non-evolutionist, Lyell made no connections of the evolutionary significance of the two sets of remains.

Thomas Huxley, eminent Darwinian and author of *Evidence as to Man's Place in Nature* (1863), surmised that the Neandertal skull represented a morphologically primitive person who was nonetheless far removed from the apes and in no sense a missing link—this opinion is what many paleoanthropologists hold today. This was also the first time the term "missing link" was introduced to explain the gaps in the fossil record. Another notable person, William King, an Irish anatomist, suggested that Neandertals represent a very primitive stage of humans and preferred to separate this extinct species from *Homo sapiens*. He therefore was the first to coin the species name—*Homo neanderthalensis* (King, 1864). Despite the species name sticking in the classification of this human group, King believed that the bones were probably of some primitive ape—or a chimpanzee-like race; and that the skull represented a "brutish" individual. He helped perpetuate the myth that the Neandertals were likely brutish, a lower kind of being; this notion has stuck to this day and will be discussed later under popular cultural representations—the caveman type. These views reinforced those stated by Rudolf Virchow, who saw the Neandertal skull as belonging to a modern human afflicted with disease and antemortem blows to the head. Therefore, William King and Rudolf Virchow are responsible for the imagery of the Neandertal as a pathological, brutish, primitive and lesser human, a notion that Marcellin Boule took a notch higher (see below).

It can be observed that further discoveries of other fossils resembling the original Neandertal began occurring across Europe. These new discoveries were to challenge the original views of the Feldhofer specimens, in that there were similarities that could not be explained by disease alone. Fossils were found in Spy, Belgium, and Krapina, Croatia. The 1891 discovery of Pithecanthropus in Java, Indonesia, also helped challenge August Mayer and Rudolf Virchow's pathological hypotheses for Neandertals, when it became clear that it was unlikely that all the widely distributed anatomical elements would belong to pathological individuals. Additional discoveries from sites in southwestern France like La Ferrassie, Le Moustier, La Chappelle-aux-Saints, and La Quina, would dispel the decades of the portrayal of Neandertals as sick individuals. But not so fast! The majority of these fossils from southwestern France were excavated between 1908 and 1911. La Ferrassie yielded two adults, along with infants and juveniles. Le Moustier in the Vézère River valley is prominent as the place that became the type site for Mousterian flint tool kits. It was the first site where the tools were found with Neandertal skeletal remains; the artifacts were described and named for that site.

To stay with the theme of this section, the discovery of the partial skeleton of a 50-year-old man from a cave in La Chappelle-aux-Saints in Corrèze, detailed in the study *Fossil Men* (Boule, 1912), is important. In this study, Marcellin Boule, a professor of paleoanthropology at the National Museum of Natural History in Paris, continued the stereotype sown by Mayer and Virchow in the early years. He actually took it a notch higher. He postulated that the Neandertals were a side branch in human evolution who went extinct without issue or were possibly wiped out by moderns represented by the Cro-Magnon stock. For imagery, Boule maliciously favored the use of the ailing Old Man from La Chappelle to create a portrayal of the Neandertals as people with stooping posture, sloping necks, bent knees, beetle brows, clumsy with inferior brains, and grasping feet like the apes. These depictions epitomized the earlier pathological hypotheses, thereby closing the debate about any fair image consideration for the Neandertals. This imagery would permeate popular writing, as well as create a lasting impact on "good" science as of 1912.

Further Discoveries

Even after the discovery and publication of the original Neandertal specimens (Tattersall, 1999; Schrenk & Müller, 2005), it took a century for the fragmentary juvenile skull, previously discovered by P. C. Schmerling in 1829–1830 in Engis Cave, Liège, Belgium, to be rediscovered and identified as belonging to the Neandertals. Before 1956, the Engis Cave material was unknown to science, just like the 1848 discovery from Forbes' Quarry on the Rock of Gibraltar. The Gibraltar skull is a well-preserved complete cranium of a female hominin. Sadly, her cranium sat unacknowledged on a storage shelf after its 1856 discovery, for several decades. As noted previously, a plethora of Neandertal specimens surfaced in many parts of Europe and West Asia, such as La Chappelle-aux-Saints, Le Moustier, La Ferrassie, and La Quina. These discoveries, made between 1908 to 1921, added skeletal elements with varying degrees of completeness. Two almost complete adult skeletons of Neandertals, which became especially noteworthy, were discovered between 1899–1905 in the Krapina Rockshelter near Zagreb, Croatia.

In 1866, the cave of Trou de La Naulette, Belgium, also yielded a fragmentary mandible, an ulna, and a metacarpal (finger bone), with some associated finds representing a mammoth, rhinoceros, and reindeer. The discoverer, Edouard Dupont, recognized the hominins as different from modern man. He observed that the toothless jaw was an apelike type in its extreme projection in the symphysis area. Other observations include the presence of very strong canine teeth and large molars that increase in size from front to back. The jaw also lacked a chin. The Naulette Man is now assigned to the Mousterian Culture (300–30 ky BP). These features were not obvious then, but Virchow dismissed them as pathological, too. Another important mandibular discovery was made in 1880 from the Sipka Cave in Moravia, in the Czech Republic. Finally, two additional adult skeletons followed in 1886 from Spy, Belgium. While these are not the only Neandertal fossils known, they have formed the core of the studies that shaped our understanding of this extinct human lineage.

PHYLOGENETIC CLASSIFICATION, NOMENCLATURE, AND NEANDERTAL ORIGINS HYPOTHESES

The tone for the formal phylogenetic placement of the Neandertals was set by King (1864) when he identified the group as a primitive group of humans and erected the taxonomic name *Homo neanderthalensis*. This nomenclature has persisted to this day. In today's paleoanthropological taxonomy, those who prefer the treatment of the Neandertal group as a species separate from us employ the term *Homo neanderthalensis*. However, some scholars see Neandertals as a subspecies of *Homo sapiens* and prefer to use our species name, *Homo sapiens neanderthalensis*, with the only distinction at the subspecies level. Thus, we would be referred to as *Homo sapiens sapiens* and they would be referred to as *Homo sapiens neanderthalensis*. These two competing phylogenetic positions are likely to continue for some time, even though more and more workers are softening their arguments for the separation of the Neandertals from us, as will be discussed further. The taxonomic debate appears to be less of a problem when compared to the more complex and variable scenarios presented for the origins of Neandertals and their relationship to other hominins.

Four competing phylogenetic hypotheses that account for the origins of Neandertals are discussed by Hublin (2009). Besides the strictly phylogenetic hypotheses for the origins and relationships of the Neandertal and other hominin groups, Dean et al. (1998) detail the origin of Neandertals by the accretion process—a slow gradation of morphological traits into what is presently the "classic" traits utilized in describing the species. The accretion model can also be discussed in light of the current origins debates (see Hublin, 2009, for a review of these debates). The first scenario for the origin of the Neandertals posits that they evolved from local (European) *Homo heidelbergensis* populations, but this entire lineage went extinct. In the second scenario, European *Homo heidelbergensis* populations gave rise to Neandertals and *Homo sapiens*. The Neandertal branch became extinct, and the modern humans flourished. Notice that the first scenario accommodates the African origins for modern Europeans, while the second model views the evolution of both Neandertal and modern humans as a local European affair.

In yet another scenario, the Neandertals are thought to have evolved from local *Homo erectus* populations. This school of thought does not recognize other hominin species such as *Homo antessessor* or *Homo heidelbergensis* that have been suggested as possible inhabitants of Europe prior to the appearance of *Homo (sapiens) neanderthalensis*. Still another scenario suggests that Neandertals are remnants of migrating *Homo rhodesiensis* from Africa. This last scenario, while plausible, does not account for the fate of the early hominins in Europe that predate the emergence of the Neandertals.

The evolution of the "specialized" morphology seen in the Neandertals has been explained by the accretion model. This model specifies that a group of Middle-to-Late Pleistocene hominins evolved in partial or complete genetic isolation from the rest of humanity—in this way, an accumulation of distinctive morphological traits evolved in Europe. Genetic drift, the change in the frequency of a gene variant (allele), in a small, isolated population, was the mechanism for this evolutionary happenstance. The effect of genetic drift is larger in small populations, and smaller in large populations. The latter part of the Pleistocene saw Europe going through glacial and interglacial climatic cycles. Neandertals are believed to have gone through a population bottleneck—a reduction in population numbers prompted by climatic fluctuations. Small isolated populations trapped in the northern climes during glacial maxima were then responsible for the founder effect conditions that led to the fixation of the classic morphological traits (extreme traits that may have been less adaptive in the long run, perhaps).

The Neandertal phase of evolution in the accretionist paradigm posits that the morphology of this group evolved gradually. Therefore, the fossil remains are often classified as either belonging to Proto- and Pre-Neandertals, Early Neandertals, the Classic Neandertals, and even more recently, the Ante-Neandertals. While the morphological descriptions can be reviewed elsewhere (e.g., Dean et al., 1998), the Proto- (or Pre-) Neandertals are believed to be represented by specimens from Mauer, Arago, Petralona, Steinheim, Swanscombe, Sima de los Huesos (SH), Atapuerca, Spain; Reilingen, Arago, and Apidima, to name a few examples.

These Pre-Neandertals appeared in Europe as early as 600–350 ky BP (peak of stage, 425–480 ky BP) and are believed to possess some Neandertal-like traits—also referred to as the "accretion phase," or gradual Neandertalization phase. The Proto-Neandertal traits are occasionally grouped in another phenetic "species," such as *Homo heidelbergensis*, or a migrant form, *Homo rhodesiensis*. The strict accretion model only recognized *Homo heidelbergensis*. The Early Neandertals, mostly dated from 300 ky BP to 130 ky BP, are represented by specimens from Krapina, Ehringsdorf, Biache, Saccopastore, Monmaurin, and La Chaise. The Classic Neandertal phase includes the specimens that possess all traits that define this group. They are often referred to as the true Neandertals. The Classic Neandertal characteristics had appeared "full blown" around 130 ky BP, but by 50 ky BP, these characteristics were disappearing in Asia. The Classic Neandertal are known from European sites such as the Neander Valley, Germany; Spy, Belgium; Forbes' Quarry, Spain; Guattari, Italy; Le Mousteir, La Quina, La Ferrassie, La Chappelle, St. Césaire, Arcy-sur-Cure, France; Kulna Cave, Czech Republic, and Vindja Cave, Croatia; and in Shanidar, Amud, and Tekish Tash in West Asia (but these West Asian materials are not included in the accretion analysis).

In the recent past, other "transitional" fossils, the so-called Ante-Neandertals, have been included in the Neandertal character list. These include the specimens from Lagar Velho, Portugal, and possibly the mandible from Tabun (Tabun C2), Israel. The transitional forms appear to show similarity between the Neandertals and the moderns and provide evidence in support of possible ties to moderns in current debates on the fate of the Neandertals. The questions as to whether the Neandertals contributed to the modern human gene pool or went extinct without issue are focussed on these specimens known to have survived until some 30,000 years BP, or as late as 27,000 years BP. At this time, the classic traits disappeared in Europe. What, then, are the classic Neandertal characteristics?

NEANDERTAL MORPHOLOGY

The morphology of the Neandertals is well known from the more than 400 fossil specimens that have been recovered over the years. These people were efficient bipedal walkers, with generally bulkier bodies than modern humans. Their physiology included larger heads and noses, dense bones, heavy muscles, and shorter stature, compared to modern humans. The pelvic bones were also wider than those of modern humans as were their brain cases (20 percent bigger than modern hominins). The stature average for the species is 5 feet and a weight of 185 pounds, with an average of 5'9" for males and 5'3" for females (Trinkaus, 1983). The lower limb bones of the arm and leg were shorter in comparison to modern humans. The stocky physiology and the wider nasal aperture are believed to be efficient and well-suited adaptations to cold environments. Because they had shorter limbs, many researchers think they must have been less agile. A detailed description of the specific anatomical character traits are discussed below. New morphological evidence from genetics will be treated separately.

Cranial Form

The overall form of the Neandertal cranium is long and relatively low, with un-thickened walls. The front of the skull preserves large brow ridges, broad noses, and jaws. In lateral profile, the skulls show a distinct occipital bun (or chignon, at the back of the head), associated with prelambdoid flattening (see parietals), giving the back of the skull a globular (*en bombe* = oval) appearance. The occipital bone morphology also depicts a suprainiac depression that is elliptical in shape, with a nearly horizontal transverse occipital torus. The mastoid process is small and shows a mastoid crest or tuberosity and an accompanying juxtamastoid crest. The base of the skull shows some flexion, but is flatter when compared to modern humans (Laitman et al., 1979). The maximum breadth of the cranium is at the midparietal level. The overall endocranial capacity is between 1245–1740 cubic centimeters (cc), compared to 1560 cc for early Anatomically Modern Humans (AMHs) and 1340 cc for living people (Holloway et al., 2004). The inner ear bones of the temporal bone shows small anterior and posterior semicircular canals with uniquely low position for the posterior canal (Hublin et al., 1996).

Some distinguishing characteristics of the upper facial skeleton include a low and receding frontal, continuous supraorbital torus—SOT (brow ridge)—arched above the orbits. A sulcus or gutter—the supraorbital sulcus—separates the frontal squama from the supraorbital tori. The frontal sinuses are generally large and restricted to the torus. The middle of the Neandertal faces shows large and rounded orbits, long and forwardly protruding mid-faces, making the nasal region appear to be been pulled forward. The nasal apertures and paranasal cavities are generally very large, with receding zygomatic arches that are not angled as in *Homo sapiens*. The maxillary bones lack a canine fossa typical of modern humans but the regions above the canine teeth are inflated. The lower faces of most Neandertals show jaws that are receding in the symphysis area, with no chins (mental osseum). The jaws show coronoid processes that are much larger than the condyloid processes and sigmoid notches that are highly asymmetrical and associated with a posteriorly projecting coronoid process (Klein, 2009: 452). The jaws also have a retromolar space—the space between the last molar and the margin of the ascending ramus (Klein, 2009: 451). This results in the combination of a long jaw, a short post-canine dentition, and relatively narrow ascending ramus. In addition, the position of the mental foramen is always below the first lower molar and single. In the internal aspect of the jaws, the mandibular foramen always shows a characteristic horizontal-oval (H-O) trait (Smith, 1978).

Dentition

Neandertals also show some unique aspects of dentition (see Klein, 2009, and references therein). The incisors (front teeth) are often large, or larger than those of earlier *Homo* and moderns. The maxillary incisors are usually shovel shaped. The maxillary and mandibular incisors usually show distinctive, rounded heavy wear on the labial (toward the lips) surface in older individuals. The premolar crowns are asymmetrical, especially the lower premolars, which have strong transverse crests with metaconids that are more mesially placed (Bailey & Lynch, 2005). The molars (cheek teeth) are smaller than in early *Homo*, but overlap with those of early, robust near-moderns and modern people in size. These cheek teeth are usually less worn than front teeth (incisor and canines); moderns show a more equal wear pattern. Taurodontism, a condition where the roots are fused and having enlarged pulp cavities, are also a common Neandertal trait (Klein, 2009: 451).

The Hyoid Bone

The Neandertal fossil record also includes one hyoid bone known from Kebara Cave in Israel (Arensburg et al., 1989). The hyoid is a small U-shaped bone that attaches to the muscles of the jaw, the larynx (voice box), and the tongue and forms the Adam's apple. This bone does not articulate with any other bone and is one of the only independently "floating" bones of the human skeletal structure. The morphology of the hyoid bone from Kebara Cave is indistinguishable from those of modern humans and may offer some answers as to the speech or language capabilities of the Neandertals.

Postcranial Form

The cervical vertebrae have long, horizontal, and robust spines at or beyond the range of modern human variation, with neural canals enlarged as in modern humans (Trinkaus, 1995b). The scapula has a glenoid fossa that is long, narrow, and relatively shallow (flat). The blade is very broad, with a deep groove or sulcus on the outer edge of the dorsal surface (Steward, 1962).

The rib cage is barrel shaped, with extraordinarily thick ribs (Gómez-Olivencia, 2009). The ribs are also weakly curved to encircle a broad, thick chest that is deep relative to the inferred stature. According to Rak and Arensburg's (1987) study of the Kebara specimen, there are some remarkable differences between the Neandertal pelvic bone and modern humans. They observed that the superior ramus of the pubic bones are lengthened and thin (also Steward, 1962). The iliac blades are also broader (laterally flared) and longer and are accompanied by a hip joint that faces more sideways. The Neandertal pelvic bones are less massive, and sexual dimorphism can be seen on greater sciatic notches (Trinkaus, 1984a). However, the female pelvic outlets are tranversely oval in shape and more similar to modern humans when compared. These observations are also corroborated by the descriptions published by Weaver & Hublin (2009).

Neandertal long bones exhibit rugged muscle and ligament attachment surfaces. The bones are robust and thickened compared to modern humans, with distal limb segments of the forearm and lower legs that are relatively short. The radii are somewhat bowed. The femuri are rounded, anteriorly bowed, with cortical thicknesses unlike those of humans. They also lack pilasters, show rugged muscle markings, and exhibit enlarged epiphyses associated with cortically thick shafts (diaphyses). The thigh bones also have low neck-shaft angle of approximately 120° when compared to the human range of 124°–135° (Trinkaus, 1993a). The Neandertal tibiae are shorter relative to the femuri, but also have rugged muscle markings, large epiphyses, with robust cortically thick shafts. The proximal shafts of the foot phalanges show strong mediolateral expansion. The pollex (thumb bone) has a distal phalanx that is roughly as long as the proximal one, and the manual phalanges have apical tufts with large rounded tips (Trinkaus, 1993b). The average stature of Neandertals based on long bones is 166 cm (5'4") compared to 178 cm (5' 8") for Anatomically Modern Humans (Trinkaus, 1983b).

THE MEANING OF NEANDERTAL SKELETAL MORPHOLOGY

In order to understand the meaning of the Neandertal biology, we can start by asking, "How do features (phenotypes) emerge?" Once we answer that question, we can then ponder the issue of explaining the meaning of Neandertal anatomy. The Classic Neandertal traits are known to have existed and peaked between 130,000 to 30,000 years BP (Hublin,1998; Hublin et al., 1995). This is a life span of about 100,000 years, representing 2,500 generations at 40 years per generation for the species or group. Evolutionarily speaking, traits are the products of genetic influence. They are the result of adaptive (i.e., natural selection) or neutral (mutation, gene flow, genetic drift) evolutionary forces that shape allele frequencies in a population over multiple generations (Weaver, 2009; also see Antón, 1994; Ackermann & Cheverud, 2004). They are also products of environmental influence, shaped by activities during life—e.g., dietary, locomotor, or manipulative behaviors. These behaviors shape the skeleton through mechanical loading patterns that are produced in a lifetime of an individual (Weaver, 2009). It can be noted that pathology is also environmental, but is treated separately below, with emphasis on modern behavior and health and life expectancy. Therefore, evolutionary and lifetime causes of skeletal forms become important in morphological analysis and help sum the biological traits that are a result of the interaction between genes and environment. The preserved Neandertal skeletal features then can be analyzed and explained in a number of ways, including; climatic adaptation, anterior dental loading, genetic drift, and pathology (see Weaver, 2009). Other specific explanations include: adaptation to cold, social behavior, life expectancy, disease, and energetics. Many of these examples of adaptive traits, especially those related to lifetime behavior, should be found in adults, but some traits under genetic control should appear in young members of the group, too. The specific biology of the Neandertals can be explained in light of the above factors.

Morphologies Associated with Adaptation to the Cold Glacial Periods

The Neandertals lived in Europe during the Ice Age and endured extremely cold temperatures in the 100,000 years of their existence. The overall stocky build, thick-set bones, large muscle markings, and short stature are due to responses to the environment following Bergmann's and Allen's rules. According to these rules, animal species that live in the hotter, equatorial climes tend to be tall and lean—an adaptation for keeping cool in the heat by way of expanded surface area for heat dissipation. Likewise in the temperate climes, it follows that the animal species respond to the need for heat conservation by way of stocky and short lower extremities (Allen, 1877: in Serrat at al., 2008). This is because such forms reduce the surface area for heat loss. The Neandertal body form therefore is a response to Bergmann's rule. The short lower extremities typically common among the Neandertals also meet the Allen's rule criteria for thermoregulation. According to Serrat and colleagues (2008), lab experiments with mice show that long term exposure to cold leads to vasoconstriction, which then affects the vascularization of the cartilaginous tissue at the end of the long bones. The continuous effect leads to thickening of the cartilage, which then pulls onto the periosteal tissue and bone, giving the bone ends the characteristic rugosity common among the Neandertal osseous remains. Of course, the accompanying adipose tissue (both subcutaneous and visceral) will provide an added advantage for heat conservation in cold environments. The subcutaneous fat tissue is found underneath the skin and surrounding muscles, while the visceral fat is common around the major organs like the heart and in the stomach, especially the omentum. These fatty tissues could also have acted as buffer to the cold and also could have provided energy during periods of food scarcity. In yet another hypothesis, Steward (2005) citing Plummer & Bishop (1994) also believes the short distal limb elements may be due to forest adaptation, as seen in African ungulates. This sort of interpretation, according to Rae and colleagues (2011), may help to explain the presence of Neandertal traits at more southerly sites not significantly affected by glacial cold.

Another aspect of Neandertal morphology associated with adaptation to the cold has been the enlarged sizes of the frontal, maxillary, ethmoidal, and paranasal sinuses. These enlarged air spaces have been reported for over 100 years as an adaptation to extreme cold temperatures (Holton & Franciscus, 2008). Holton and Franciscus (2008) and those before them proposed that enlarged paranasal sinus were meant to warm air before going to the lungs. The mechanism of this supposition is now under scrutiny. According to Rae and colleagues (2011), the opposite is found to be true in regard to the highly pneumatized Neandertal cranium, whereby the frontal, maxillary, and paranasal sinuses are enlarged. A Rae et al. (2011) study found that animals in the wild and in laboratory conditions show remarkable reduction in the volume of the sinuses in extreme cold conditions and that the very large sinus conditions recorded among Neandertals are actually due to scaling—that is, they are larger because Neandertals are overall larger than modern humans when their relative cranial sizes are compared. This finding helps to rule out the claims for these cranial morphologies among Neandertals and that only postcranial distal limb morphologies likely are due to extreme cold adaptations.

The Neandertal bulky physique has also been explained as due to strenuous physical activities and a hard lifestyle that led to stress in both the muscle and bone. The hunter-gatherer lifestyle likely exposed this human group to a lifestyle that required their physiology to adjust to hard life. A good example is the cortical thickening of bones (Odwak, 2000; Cowgill, 2007). This physiological adjustment would require the mechanical loading by weight-bearing tissues to function properly (Ruff, 2000). While physicality, a lifetime behavior, may have contributed to the type of physique the Neandertals had, it does not explain why young individuals of the species have the same physiology (Golovanova et al., 1999). This may imply that the presence of large muscles and thickened bones were conferred through genetics and also were likely a species-wide response to Bergmann's rule.

Comparisons of the bone rugosity between males and females of this human group also show no clear differences, except that the two sexes are still sexually dimorphic in the modern human pattern (Trinkaus, 1980). This may be interpreted as a sign that there was no division of labor among the Neandertals because females would likely show gracile forms, due to the expectation that they carry out less strenuous work (see a discussion in gender division of labor among hunter-gatherers in Marlowe, 2005). Another reading into this invariant robusticity among both genders is that if Neandertal females participated in strenuous labor just like their male counterparts, they would have reduced birth rates and low survival rates of their offspring. This could have a direct impact on their success as a species. The high death rates and the number of skeletal lesions among the females of the group also indicate that they were equally living dangerously and suffered trauma and death (Trinkaus, 1978, 1985, 1995; 2012). These lines of evidence show us that the Neandertals endured hard life to survive the Ice Age, but still do not give us satisfactory answers as to their physiology. The most parsimonious conclusion would be that both genes and environment resulted in the physiology of the Neandertals.

What do the Skull and Brain Sizes Tell us About the Neandertals?

The sizes of the Neandertals' skulls and brains are large, with the endocranial volume topping 1450 cc, compared to modern humans, with an average of 1340 cc (Holloway et al., 2004). In the case of the Neandertals, their brain volumes were approximately 20 percent larger than modern human brains, but proportionally, Neandertal brains were smaller, especially on the frontal region. The back of the brain—the occipital lobe, where the centers for sight and touch are situated—is, however, well developed. The front of the brain, the frontal lobe responsible for a number of functions, including speech, memory, and analytical reasoning, is, on the other hand, relatively small compared to modern humans but not in any way different from their contemporaries. Based on the reconstruction of the endocranial casts, the left side of the brain appears to be larger than the right side (Holloway et al., 2004) implying that Neandertals were mostly right-handed if these brains were lateralized and functioned in the patterns seen in modern humans.

The brain sizes of Neandertals are larger than those of modern humans. This fact has led to many studies concerning birthing and life history patterns between modern humans and other hominins (Trinkaus, 1983a; Ponce de León et al., 2008; Weaver & Hublin, 2009). The brain sizes of babies at birth have been estimated to correlate with pelvic outlets of Neandertal females. Because of the elongated pubic ramus and modest pubic inlet results shown among Neandertals, Trinkaus (1983a) proposed that Neandertal babies had precocious brain development—that is, the brains achieved a sizable degree of growth in utero, for Neandertals to achieve the high cranial volumes. Other arguments also imply that Neandertals had a shorter childhood, that their children had less time to develop cognitive abilities in the way of learning in comparison to other human groups. It is known that modern human children achieve the majority of their brain growth extra utero compared to living great apes (Vinicius, 2005). If precocious development or some faster development of the brain relative to modern humans is true (but see brain growth and pelvic size below), this argument may likely exaggerate the learning differences between Neandertals and the moderns. The relative percent difference in brain sizes and supposed short childhood may have only a minimal effect on Neandertal learning. This argument does also conflate the high rate of childhood mortality among this maligned human group with the lack, or deficient level, of learning. While high rates of infant mortality were likely common in all human groups in prehistory until the recent past, all the Neandertal children who survived to adulthood probably acquired adequate knowledge for surviving as responsible and properly functioning members of the society. How can we account for the celebrated achievements of the Neandertals? Examples are numerous; they made the Mousterian tools, possibly the Châtelperronian ones as well; they buried their dead, took care of their sick and elderly, had forethought, and probably had a working language. These achievements are not likely to be corroborated by the precocious-brain-development/lack-of-proper-learning scenario.

Did Neandertals Show Some Cognition, Forethought, or Planning?

The answer to this question is yes. Neandertals left behind various forms of evidence to attest to this fact. For instance, they survived for over 100,000 years in western Europe when the climate was intolerant. They also buried their dead. They have the first evidence of indisputable burials in human history (Pettitt, 2002). Some workers have asked if Neandertal burials were ritualistic or simply actions to avoid attracting scavengers. Neandertal burials usually occurred inside caves/rock shelters. Some were filled with burial goods and pollens that have been viewed as offerings, or at least interpreted as such, as likely presence of flowers that accompanied the dead (Solecki, 1975). The questions of whether the burials were intentional or not should not come up in the first place (see Chase and Dibble, 1987). For any burial to occur, there must be planning. The decisions include the location for burying the remains of the dead member of the community, the depth of the grave, what goods would accompany the individual (Vandermeersch, 1970), how the body is interred—in a flexed or prone position? (in Harrold, 1980). How many people are digging the grave? What tools are used to dig, are tools available, and how hard is the soil? These are some likely decisions indicating a presence of forethought. Many workers agree that Neandertals intentionally buried their dead (Pettitt, 2002). These decisions reflect a group commitment to respect their dead and possibly a belief in an afterlife. Above all, there is the presence of forethought, which is processed at the frontal lobe. Many arguments for "less intelligent" Neandertals lack comparison between Neandertals and their contemporaries, both in Europe and in other parts of the Old World. For instance, it would be useful to know how modern humans fared in solving similar problems in Ethiopia, for example. There were early modern humans living in the Omo Valley in Ethiopia around 190,000 years before present, and they also largely show small frontal lobe development. Did they have forethought?

What can we Learn from Teeth?

The Neandertal front teeth, the incisors, are shovel shaped and show a high incidence of attrition on the labial sides, the areas toward the lips (Trinkaus, 1983b). This wear pattern is attributed to possible use of the front teeth to hold objects (Trinkaus, 1983b; Rak, 1986). Such occurrences are common among many adult-age individuals and in modern Inuits (Brace et al., 1981), where the front teeth have been used as a fifth appendage. The wear of this pattern is therefore not due to masticatory uses, but rather a behavior more in tune with a cultural pattern.

Did the Neandertals have a Language?

The debate about whether the Neandertals had a language or not has been around for some time. One school of thought maintains that Neandertals did not have the capacity for speech or language as we know it in modern humans (e.g., Lieberman & Crelin, 1971). Another believes that Neandertals were equally capable of speech and language just like us (Arensburg et al., 1989). The evidence from a 60,000-year-old hyoid bone from the Kebara cave (mentioned above) is most positive. The hyoid, and more so its position in the neck, allows humans to produce a wide range of sounds necessary for language. The human hyoid is much lower in the neck, allowing us to do exactly that—produce a wide range of sounds. However, anthropologists who claim that the Neandertals did not have a language or that their language was primitive think that the position of the hyoid bone among the Neandertals was much higher, thereby limiting the number of speech sounds they could make (Lieberman & Crelin, 1971). This hypothesis relies on the calculation that chimpanzees and other apes, including most mammals, when compared to modern humans show less flexion of the base of the skull. The flattened basicranium correlates with the much higher position of the hyoid bone among these animals, allowing them to swallow and breathe at the same time. As we know, apes and other mammals are only capable of a limited range of phones (language sounds). Lieberman and Crelin (1971) also erroneously compared the reconstructed La Chappelle Neandertal neck region to those of an infant human and a chimpanzee, again using the Welcker angle, which was not replicated in their chimp-human neonate study. This led to an erroneous conclusion about the Neandertal language. Falk (1975) challenged the use of apes and neonates with the remains of a fully adult Neandertal. The 1989 reconstruction of the famous La Chappelle fossil by Jean-Louis Heim of the National Museum of Natural History, France, shows the presence of a slightly flexed basicranium in Neandertals, reminiscent of modern humans. This evidence supports the fact that Neandertals had the hyoid and larynx lower in the throat compared to apes, supporting the conclusion that they could also speak as modern humans.

Those who believe Neandertals had a language and could speak like us rely on the anatomy of the hyoid and other evidence. According to this school of thought, the hyoid found at Kebara is impossible to tell apart from those of humans in many aspects (Arensburg et al., 1989). The close similarity of Neandertals and modern humans in this respect implies that the organs associated with this bone may have functioned in similar ways. The slightly flexed base of the Neandertal skulls compared to humans may have had very little impact on the range of sounds they could produce. Falk (1975) also reiterates that the descent and positioning of the larynx and the hyoid are produced in relation to bipedality, a quality lacking in chimps and human neonates. The positioning of the hyoid lower in the neck readily predisposes us to choke when we attempt to breathe and swallow at the same time. The fact that this potentially disadvantageous trait survived natural selection suggests that the evolutionary advantage of speech was very important, and full bipeds like Neandertals shared the benefits of speech and its potential hazards.

Besides the evidence from the hyoid bone, the proponents of pro-speech for Neandertals also look at the brain. In modern humans, the well-known speech centers include the Wernicke's and the Broca's areas. The evidence from Neandertal neurocranial endocasts shows that these two areas important for speech are actually well developed among the Neandertals. Assuming that these important brain centers performed similar functions as humans, it is safe to conclude that Neandertals had the capacity for language. The other line of evidence for Neandertal language comes from the emerging genetic evidence (discussed later). The language gene (FOXP2) has been found in the DNA of Neandertals and has been shown to be identical (Krause, 2007). That means the separation between the Neandertals and early modern humans probably happened after the mutation of this gene as we know it in their common ancestor more that 130,000 years ago.

In the archaeological record, the presence of art, including cave art, may also signify the capacity for high cognition and the ability to understand complex language. Other evidence of indisputable burials, complete with grave goods, may also help support the claims that Neandertals were cognitively advanced like us—including their linguistic capabilities.

What can we Learn from Other Neandertal Postcrania?

The pelvic remains of Neandertals are now known from the two Israeli localities of Kebara and Tuban. These remains have helped in the interpretation of the implications of the pelvic dimensions in locomotor, obstetric, and life history (Trinkaus 1983a; Ponce de León et al., 2008; Weaver & Hublin, 2009). The hip joint that is more sideways-facing may imply that the Neandertal bipedal locomotion was different from that of modern humans. The locomotor variation suggested in this interpretation may not mean that they were not efficient bipedals, and the exact nature of that difference has not been clarified. The obstetric implication of the pelvic bone has solicited some debate. The forward expansion or projection of the pubic rami, the flaring iliac blade, and a slightly larger spheroidal pelvic inlet can be interpreted to mean that Neandertal females probably gave birth to large babies with more brain volume. As mentioned previously, this interpretation meant that the brain of Neandertal babies likely achieved larger volumes during growth in utero and were equally larger at birth than those of humans. This suggests a precocious "maturity" of brain development. Another reading into this larger pubic ramus and slightly larger pelvic outlet is that the gestation period of the Neandertals was likely longer or that the females had reduced birthing constraints. After the discovery of some juveniles or neonates from Mezmaiskaya Cave in Russia (see Ponce de León et al., 2008), the initial interpretations of these birthing and brain size difference between Neandertal and modern humans have changed. The Neandertal brain size at birth is now thought to be similar to that of *Homo sapiens* and that Neandertals were likely facing similar obstetric challenges (Ponce de León et al., 2008; Weaver & Hublin, 2009). Ponce de León and colleagues, however, add that the brain growth rate could have been higher during early infancy, although this pattern only resulted in larger adult brain size mentioned earlier, but not in earlier completion of brain growth. This new insight clarifies the notion held by earlier studies, where there was the notion of precocious brain growth and its attendant shortened childhood and learning time for the Neandertal children. Weaver and Hublin also still maintain that the Tabun pelvis canal shape still indicates that Neandertals may have maintained some primitive birthing pattern and that the rotational birth pattern may have developed later in human evolution.

WHAT DOES THE PRESENCE OF LESIONS ON NEANDERTAL SKELETAL REMAINS MEAN?

The initial interpretation of the Neander Valley fossils by August Mayer was that the specimen represented a pathological case of some ancient humans. Later, in 1872, Rudolf Virchow supported Mayer's interpretation of the fossils. This misevaluation of the Neandertal skeleton rendered them as pathological from then on. The original specimen had a healed fracture on the left elbow, however. The perpetuation of this "pathological insight" from the discovery

BOX 1: SOME HEALTH PROFILES OF NEANDERTALS

Specimen	Trauma/lesion type and element
Neanderthal 1	– fracture below the left elbow
La Chappelle-aux- Saints	– broken rib, edentulous, vertebral osteoarthritis
La Ferrassie 1	– injury to the right hip
La Quina 5	– lesion on right upper arm
Sala	– lesion on right forehead
Krapina	– broken arm, never rejoined
Shanidar 1	– healed fractures of two arm bones, one atrophied with pathological distal end; the right ankle and big toe show arthritis, fracture of 5th instep bone
Shanidar 3	– partially healed sharp object wound on the 9th left rib; right ankle and foot joints suffered arthritis
Shanidar 4	– show healed broken rib
Shanidar 5	– trauma (blow) to the left of the forehead

Sources: Trinkaus (1978); Dawson & Trinkaus (1997)

of the initial Neandertal specimen would then set the pattern for how the development and evolution of the views about this maligned species would continue. These views permeated all aspects of the paleoanthropology of Neandertals, including their phylogeny, behavior, health, and life—in a process of "othering." Some of these issues are addressed in this section. Many Neandertal fossils actually do show evidence of trauma and skeletal lesions (Box 1). This bring us to the initial question for this section: Just what does the presence of skeletal pathologies tell us about these extinct hominins?

First and foremost, it is important to point out that the presence of lesions in a skeleton means that the individual was affected by some form of injury or disease. In the case of injury due to fracture, for example, the bone will resorb by building new bone on the surface of the wound. This process continues until healing is completed. This process of healing requires time and support for the sick for healing to occur. The presence of healed scars on bones means the injured individual survived the trauma. This brings us to the second point.

That is, do skeletal lesions indicate poor health or good health? This answer represents a health-disease paradox. The argument can go thus: if an individual has been afflicted by a disease or trauma, he or she can die or heal. If he or she dies early due to the trauma or illness, there may not be any sign left on the bones resulting from the disease. However, if they suffer but survive the pathology by healing, there will or may be evidence of the effect of disease or trauma on the bone. This means that the presence of lesions in a bone may actually imply better health. It shows that the sick survived the disease. This presents a paradox in the definition of health as the absence of disease. In these terms, the many survivors of these disease indications among the numerous Neandertal remains may mean that the Neandertals were "healthy" as opposed to sick. Take the case of Shanidar 1, for instance—the individual suffered from many injuries during life, some of which may be related. The right forehead shows signs of cuts on the scalp. The left eye socket manifests trauma, possibly due to a blow that crushed the orbits and may have caused lifelong damage to the left eye and possibly the brain. If this scenario is true, then the damage to the right side of the brain may have caused poor innervation and vascularization of the right side. This Neandertal man could have been paralyzed on one side of his body during life. The right side of the body suffered from an atrophied shaft of the upper limb and arthritis of the ankle and big toes. The left arm bone has normal size but also shows a healed fracture. The atrophied arm bone may also have been due to an injury that led to infection or intentional amputation. Additionally, the Shanidar 1 individual has a fracture on the fifth instep bone. The survival of this individual with these ailments, when he was aged between 40 and 60 years and probably not reproductive to the group, may be interpreted in various ways.

Another case worthy of mention is that of the Old Man from La Chappelle, who died at an advanced age of 50 years or more. The Old Man from La Chappelle may have suffered a broken rib in addition to being edentulous (lacking teeth). The latter situation would imply that for him to survive, members of his group probably chewed food for him in order to feed him or they had improved methods of cooking that enabled the elderly to feed on softer food. The sanguine view would be that the Neandertals represent a population who had human compassion and cared for their sick and the elderly. The multiple healed and unhealed lesions certainly imply that life was hard and dangerous among these groups of humans. These characteristics—care for the sick, the old, and possibly cooking skills—can be read as evidence for humanness and are used to describe humanity today. The conclusion can therefore be made that the Neandertals shared the qualities of humanity with us.

Having said this, the following question may be of significance. What caused the skeletal lesions, or what do the lesions tell us about Neandertal life? Erik Trinkaus (1978; 1995) has postulated that the Neandertals were prone to dangerous living. Berger and Trinkaus (1995) compared Neandertal health based solely on fractures to that of Texan rodeo riders. In the current human population, rodeo riders are the only group showing very high incidences of trauma by fractures. Based on the archaeological record, the evidence from their middens shows that they hunted large game, mostly several times larger than they were. They also probably used short spears for stabbing. Their short stature also indicates that they had short extremities relative to their torso and that their upper bodies were more muscular compared to their lower extremities. These lines of evidence have led to the interpretation that the Neandertals used a close-quarter stabbing (forces) during large game hunting. In these instances, they would have been in close contact with large game and easily trampled. This explains the high incidence of bone trauma and possibly the high death rate at the prime ages of 20 to 30 years old (Trinkaus, 1995). Death at old age was rare, as depicted in their life expectancy profiles. It means the youth were more active and were involved in these encounters that endangered their lives or caused deaths. It is likely that many accidents of this nature may have led to deaths due to infections. Other possible explanations may be taphonomic or differential interment of people of "adult" age. The Turkana people of northern Kenya consider only people of reproductive age as worthy of consideration for a proper burial, and not children. These ethnographic factors may also have played a role in the representation of skeletal material in prehistoric times. However, it can be pointed out quickly that Trinkaus (1995) included data from many prehistoric groups and some hunter-gatherers, but showed clearly that the mortality patterns were fairly similar.

Other evidence has shown the presence of a partially healed sharp object trauma on the 9th rib of the Shanidar 3 individual (Trinkaus, 1978). There is a likehood that a spear (with arrowhead) was used in close-quarter stabbing or jabbing by a right-handed individual to inflict this wound. This injury and its position on the 9th rib may have caused the left lung to collapse, causing death to the victim, probably days later. This evidence has been used as indication by those who hypothesize that violence or war between Neandertals and modern humans likely caused the extinction of the Neandertals. It can be suggested just as easily here that such evidence is not conclusive. While the presence of sharp object trauma to the left of the chest and the presence of lesions on the ribs is undisputable, the way that arrow or spear point happened to inflict that wound can be argued variously. For one thing, it is possible, as proposed by some researchers, that it is evidence of the decimation of the Neandertals through war. It is also possible that this could be Neandertal versus Neandertal violence, just as modern humans also fight when peaceful means of solving disputes fail. Better still, the incident just as likely could have been an accidental occurrence. Either a member was aiming at an animal prey in close quarters and inadvertently jabbed his hunting party member. Because all these scenarios are likely, it is impossible to pick one as absolute truth without further evidence or a well-documented pattern. This scant evidence still remains the only case of undisputed violence among Neandertals. With all the pathological evidence, we may be curious to know the life expectancy of the Neandertals.

NEANDERTAL MORTALITY PROFILES AND LIFE EXPECTANCIES: HOW LONG DID THEY LIVE?

The mortality profile of available Neandertal fossils represents individuals of all ages, from infancy to old age (Trinkaus, 1995). The human osteological methods are used to estimate the developmental age at the death of an individual. The methods are based on cranial suture fusion stage, dental enamel development and eruption schedule, long bone length and epiphyseal fusion stages, pelvic features of the symphysis, and the auricular facet and other degenerative conditions like dental wear, among others. Some of the skeletal remains also help estimate the sex of an individual. Some features of the skull and pelvis, in addition to some postcranial bones, can be utilized in the estimation of maleness or femaleness. For fossil hominins, both ape and modern human growth schedules are used to help estimate the patterns seen in early hominins. Based on these studies, Box 2 shows the ages of particular fossil examples.

According to Trinkaus (1995), the longevity data show that immature mortality distributions were within the ranges of the recent human samples. There was also a high mortality for young prime-age adults and low older-adult mortality.

NEANDERTAL GENETICS

The advancement in the science of genetics, especially in the methods of extraction and copying of ancient DNA (aDNA) material, has helped us understand more about additional Neandertal biological makeup. The study of functional DNA of the Neandertals is still ongoing, but some successes are reported in this section. Many fossil specimens discovered so far have shed light on important anatomical forms and function and provide important phylogenetic information. When it comes to genetic evidence from DNA, scientists have only been lucky in

BOX 2: EXAMPLES OF SKELETAL AGE AT DEATH AMONG NEANDERTALS

Specimens	Estimated skeletal age at death
Amud 3	- 4-year-old
Châteauneuf 1	- 3-year-old
Gibraltar 2	- 3.9-year-old
Kebararemains	- 6- to 12-year-olds
Kiik-Koba 2	- 1 year
La Chappelle-aux-Saints	- Old Man, 45–55 years old
La Ferrassie	- 7-to-9-month-old infants
La Ferrassie 4	- fetus
La Ferrassie 4a	- neonate
Lagar Velho	- 4-year-old juvenile
Neander Valley	- approx. 60 years old
Pinar 7	- 6–7-year-old
Subalyuk 2	- 3–4-year-old
Shanidar—adults	- 40 to 60 years old
Shanidar 6 & 8	- infants, under one year of age
Shanidar 1	- 35–50-year-old
Tekish-Tash	- 12-year-old

Source: Trinkaus (1978; 1995)

obtaining some DNA material from a few specimens representing various ages and regions. These specimens include: the femur from Vindija Cave, Croatia; two long bone fragments from El Sidrón, Spain; the original Neandertal from the Feldhofer grotto, Germany; and a cranial fragment from Mezzena Rockshelter, Monti Lessini, Italy, dated to 38,000, 43,000, 45,000, and 50,000 years BP, respectively. The number of DNA analyses has increased rapidly in the recent past. The genome studies mostly involve mitochondrial DNA (mtDNA), a small circle of DNA found outside the nucleus in the mitochondria, the organelle that provides most of the energy to run the cell. This DNA is found in larger amounts than nuclear DNA, and it is tougher—as a circle, it has no free ends to attack. Mitochondrial DNA is inherited strictly through the mother. Other nuclear DNA sequences have also been successfully extracted. Nuclear DNA is inherited from both parents. The following paragraphs discuss the significance of a number of functional genes that have been identified from the DNA extracted so far.

The Neandertal genes known today include: the FOXP2, MC1R, TAS2R38, ABO blood group genes, and the Microcephalin haplogroup D genes (not treated further here). The FOXP2 gene was extracted from a femur of the 38,000-year-old male fossil from Vindija Cave, Croatia. Other evidence of the FOXP2 gene came from the two long bone fragments (numbered 1253 and 1351c) belonging to mature males from El Sidrón in northern Spain. FOXP2 is a regulatory gene—that is, it controls the expression of other genes. It is especially important in brain growth and development at the embryonic stage. FOXP2 also codes for morphological expression, especially on the growth and form of the facial muscles, tongue, and some neck muscles. It has been suggested that FOXP2 may also be responsible for the human cognitive ability to understand complex language processes (Lai et al., 2001). Therefore, it is simply referred to as the language gene. The inactivation of one allele of FOXP2 results in the conditions similar to adult-onset Broca's aphasia, whereby patients have problems in orofacial movements and language processing (Lai et al., 2001; Enard et al., 2002; Vargha-Khadem et al., 2005). It is important to note that the function is more complex, that it may not directly influence language processes, or the gene may not have functioned in similar fashion in Neandertals, but this is unlikely.

The FOXP2 gene in humans shows two mutations that separate us from other primates including the chimpanzees (Enard et al., 2002; Krause et al., 2007). Neandertal DNA studied also shows similar patterns of mutations identical to that of modern humans (Krause et al., 2007). This mutation as we know it in humans probably happened over 200,000 years ago (Lai et al., 2001; Enard et al., 2002), but the similarity between modern human FOXP2 and that of Neandertals leads to second-guessing these dates. What are the implications of this finding in understanding Neandertal life? Does FOXP2 really tell us anything about language evolution? Or does the mutation of FOXP2 disrupt codes for morphology? These questions are complex to answer, but if the gene is responsible for human speech and language along with other functions, then we can, with some confidence, infer that the gene performed similar functions in Neandertals. The early mutation date may also mean that the Neandertals may have "branched" from the human line when the gene was already present in its current form. The so-called early modern humans also coexisted with Neandertals for more than 100,000 years. This makes it likely that they shared the FOXP2 gene through interbreeding—see the section on extinction for further debate on this subject. More recent work suggests that since the mutations of FOXP2 in Neandertals are similar to those in modern humans, the recent dates for this mutation in humans may likely reflect estimates based on data of extant human diversity. This underscores the need for the direct use of Neandertal DNA sequences (see Trinkaus, 2007). In this case, the mutation reveals that Neandertals acquired this gene before they diverged from the early archaic human stock, probably 400,000 years ago (Green et al., 2010) and not likely from gene flow or interbreeding. The absence of the two nucleotide substitutions in the ape FOXP2 genes (Krause et al., 2007) may also mean that the derived state is a condition that happened right at the time of our separation with the apes over five million years ago.

The Melanocortin 1 receptor (MC1R) gene has now been synthesized in Neandertals from Monti Lessini, Italy, and El Sidrón, Spain (Lalueza-Fox et al., 2007). In humans and other vertebrates, the MC1R gene is known to regulate pigmentations that affect both skin and hair color. People of European origin with pale skin color and red hair carry a variant of MC1R with reduced function. Lalueza-Fox and colleagues' data show that an inactive MC1R variant occurs in Neandertals, with an A-to-G base substitution in position 919 of the gene. This single base substitution makes the gene slightly different from the modern human variant of the same gene, implying that there was an independent evolution of the gene. However, the expressions of the gene (pale skin and red hair) are suggested to have been similar to those of modern humans (Lalueza-Fox et al., 2007) of European stock. While this may be used to argue that Neandertals and modern humans are different, it is not conclusive in differentiating the two groups. Other examples of independent mutations are common in modern humans. For example, the mutations in lactase genes among human groups who domesticate milk-producing mammals are known to have occurred independently in northern Europe, East Africa, and the Arabian Peninsula (Gluckman et al., 2009). These groups do digest the milk energy, lactose, and are therefore lactose tolerant, but certainly cannot be considered different species of humans. The Neandertal variant of the MC1R gene suggests that they were red-haired and possibly pale-skinned. For starters, it can be argued that Neandertals probably had various genes that control for pigmentation. It may be that the genes came from red-haired individuals, and this evidence is not enough to think of all Neandertals as red-haired people. Again, the expression of colors of the hair, eye, skin and other melanin-bearing tissues are under the control of more than one gene—thus, traits under polygenic control. Genetic analysis shows that the Neandertal variant of MC1R probably evolved independently from that of modern humans. What we probably will never know in this scenario is how the other humans who lived elsewhere compared. Further research may clarify these questions in the future.

As with pigment genes, another recent study by Lalueza-Fox and colleagues in 2008 including the El Sidrón fossils also suggest that the O blood group (O01 haplotype) in humans predates the human Neandertal divergence. Blood group O is the most derived blood group type that evolved from the blood group B. ABO blood groups are controlled by the same gene locus. Other apes, such as gorillas, show a B-type blood antigen, but it is not an exact copy of the human blood. Chimpanzees carry two alleles that correspond to the human A and O, but both apes show species-specific mutations for their variants of the blood types. Now, the observation of the Neandertal O01 haplotype is similar to humans in that the appearance of the haplotype was early, before humans and Neandertals separated over 500,000 years BP.

The bitter taste perception (TAS2R38) gene, also know as the PTC gene, has also been isolated from Neandertals from El Sidrón (Lalueza-Fox et al., 2009). This gene was isolated in 2003. The expression of this gene allows those who have an active form to taste phenylthiocarbomide (PTC), a toxic molecule in many plants. The sense of bitter taste protects us from ingesting toxic substances present in some vegetables. These toxins can affect the thyroid when ingested in large quantities (Lalueza et al., 2009). This gene is expressed in nearly all plant-eating animals and is important in "decision making" about which plants are edible and which ones are toxic or poisonous for consumption. In humans, this gene was important, especially in our prehistoric past, where most accessible plant food items may have required the need to choose those that are edible—that is, those with low levels of PTC toxin. It must have been selected early in our primate lineages when our primitive primate ancestors included plant matter in their diet. Because of domestication, man has reduced toxicity in many plant crops, mainly through artificial selection and hybridization. Many humans today are strong tasters early on in life, but in adulthood, the gene expresses variously. Some people are strong tasters, others are weak or nontasters. If Neandertals share this gene with modern humans, it shows our close relationship. The presence of this gene implies that bitter taste perception and its derived non-tasters predates the divergence of lineages leading to Neandertals and modern humans. Neandertals and modern or archaic humans had their last common ancestor some 450,000 years ago (Green et al., 2010), long before the encounters in western Europe.

More recent data comparing humans and apes suggest that humans and chimpanzees are 98.77 percent similar. The emerging comparisons between modern humans and Neandertals suggest a 99.7 percent similarity in genome makeup. So, are Neandertals different from us? Are they an entirely different species? Was there possible interbreeding between them and humans? Do modern people carry any DNA shared by Neandertals? Well, from the above discussions, we share a lot in terms of genes with Neandertals. More of these issues can be treated further in the section dealing with the Neandertal extinction question.

NEANDERTAL BEHAVIOR AND THE ARCHAEOLOGICAL RECORD

During the course of hominin evolution, three major lithic divisions, or industrial complexes, have been observed. These can be divided into the Lower, the Middle, and the Upper Paleolithic industries in reference to European archaeological nomenclature. In Africa, the archaeological assemblage divisions include the Early, the Middle, and the Late Stone Ages. The differentiation between the cultural traditions are based on techniques of manufacturing, raw materials used, tool kit assemblage, and sometimes with human species association. The Lower Paleolithic, dated from 1.7 mya to 300,000 years BP, is mainly associated with archaic Homo and *Homo erectus* (Gamble, 1993; Lewin, 1998). The industry includes pebble tools, and later periods saw the introduction of the Acheulean tools such as hand axes. The African equivalent of this industry, the Early Stone Age, dates to 2.5 Mya (Semaw et al., 1997). In the Middle to Late Pleistocene, a number of lithic industries emerged both in Europe and Africa. The European record at this time includes industries divided into Middle and Upper Paleolithic.

The Middle Paleolithic industry dating from around 300,000 to 42,000 years BP is associated with most archaic Homo groups, including the Neandertals. Neandertals are mainly associated with both the Mousterian and Châtelperronian tool traditions. The Mousterian is named after the initial finds discovered in 1863 in Le Moustier, France, where the first association of this artifact industry was made (Klein, 2009: 481). This period was composed of the Mousterian industrial tool complex with its Levallois toolmaking techniques, which started becoming more prominent after 200,000 years BP and became more refined with time. It is important to note that the word Mousterian is sometimes loosely used for similarly produced assemblages that are not always associated with the Neandertals. The Mousterian cultures extended across Europe and North Africa and included tools such as points, scrapers, and Levallois flakes. It is important to note that there was overlap between the Mousterian cultures and those of the Upper Paleolithic. Stone tool implements, including blade tools, signaled the beginning of the transition to Upper Paleolithic with the appearance of tools such as burins and notched and retouched tools. Earlier on, the Neandertal assemblages showed 40 percent hand axes, and later assemblages saw reduction in hand axes to a mere 8 percent. The increase in the number and variety of flake tools and ultimate de-emphasis of the handaxes mark this period. These later changes in the composition of the assemblage, including more refined tool production techniques, defined the Upper Paleolithic. At this time period, bone tools, musical instruments, cave paintings, symbolic artifacts, and decorated items used as personal ornaments became prominent.

Around 42,000 years ago, other lithic production techniques or traditions also emerged, known as the Upper Paleolithic traditions. These include: the Châtelperronian (the earliest industry in the Upper Paleolithic, known between 42–32 kya); the Aurignacian (begun 40 or 29 kya); the Gravettian (begun 21,000 years ago), and the Magdalenian-Solutrean traditions or cultures (begun 17,000 years ago) (see reviews in d'Errico et al., 1998; Klein, 2009). The period after 15,000 years ago ushered in several tool production techniques or cultures and the beginning of the Neolithic cultures and the so-called "agricultural revolution." The transition to European Upper Paleolithic—which may have started as early as 60,000 or 65,000 years ago—with the attendant industrial complexes has been read as marking a kind of "revolution" by some archaeologists, with some noting the association of the improved techniques to modern humans (Klein, 2009). It has been suggested that modern humans went through a rewiring of the "cerebral rubicon" (Gamble, 1993)—in other words, a period of improved brain function signaled in the new artifact types. This so-called explosion of behavioral modernity is claimed to be a hallmark of humanity—this is when we became humans (Gamble, 1993; Klein, 2009). The period marked a speciation event and the evolution of modern humans as we know them today. Therefore, Neandertal contribution to any of the more improved Châtelperronian artifacts is disputed on these grounds, as they were not like us. The opposing views come from the African archaeological record, where the signals for behavioral modernity seen in Europe already started appearing earlier in Africa, albeit gradually as represented by the artifact record from the Kapthurin in Kenya (McBrearty & Brooks, 2000), Katanda in Congo (Yellen et al., 2001), and some cave sites in South Africa (Henshilwood et al., 2002; d'Errico et al., 2005; Mourre et al., 2010). If the African record in the appearance of symbolic items, bone tools, advanced flake tools, and decorated items is true, then it is argued there was no sudden revolution in human capacity (sensu McBrearty & Brooks, 2000). The alternative reading of the European record can be that modern humans migrated into Europe intensely after 60,000 years ago.

The association of Neandertals with much of the terminal Mousterian tradition is undisputed. However, the Upper Paleolithic traditions such as the Châtelperronian and the fairly overlapping Aurignacian preceding that period are marked by controversy (Hublin et al., 1996; d'Errico et al., 1998; Flein, 2009; Higham et al., 2010). The controversy stems from a number of reasons. One issue is that Aurignacian industry is associated with modern humans who migrated from Africa into Europe, and the assemblage, composed of bone tools, jewelry, and refined lithic elements, are believed to be made by modern humans and therefore signal "behavioral modernity."(Klein, 2009; Higham et al., 2010). The appearance of these cultural items in largely Neandertal sites or levels within archaeological sites (especially at Arcy-sur-Cure, France) are read to mean artifact intrusion, geological disturbance from the Aurignacian levels, or poor archaeological techniques (Higham et al., 2010). On the other hand, two fossil bones belonging to Neandertals are known from Saint-Césaire and Arcy-sur Cure and are associated with the Châtelperronian industry (Hublin et al., 1996). Even though the fossils squarely put Neandertals as the makers of the Châtelperronian—showing artifacts made from bones, ivory, and decorated ornaments—some archaeologists interpret the data to mean that the Neandertals were not the makers of Châtelperronian. That Neandertals were imitators of humans after the contact, implying that they were technologically "inferior" (e.g., Hublin et al., 1996).

Other researchers do not take well the portrayal of Neandertals as lacking in the ability to make symbolic art, including the bone or ivory tools. These pro-Neandertal studies interpret the data to mean that Neandertals were progressive in their artifact techniques and are the makers of the Châtelperronian—even the previous evolved Mousterian. They also assert that the claim that the upper Aurignacian levels may have been due to intrusion or disturbance is wrong. For instance, d'Errico and colleagues (1998) show that at Arcy-sur-Cure, three levels, X, IX, and VIII, are all Châtelperronian, with the lowest stratigraphy (level X) actually showing more bone and other artifact counts to be higher than the next two levels, IX and VIII. There is also a difference in Level VIII artifact density compared to the immediately overlying Aurignacian level VII above it. The interpretation of this scenario is that if Aurignacian artifacts are due to disturbances (as in Higham et al., 2010), they should be more abundant in the immediately adjacent level VIII than level X (d'Errico et al., 1998). But this is not the case. The conclusion is that Neandertals are the manufacturers of Châtelperronian, with fairly unique grooving techniques of bone. The imitation or acculturation hypothesis is perceived as a pursuit to demean the intelligence of the Neandertals, as has been the case for many years since their discovery.

Those who think Neandertals are a separate species tend to downplay the achievements made by the Neandertals in their artifact production, often propounding the notion that the techniques were inferior compared to those of *Homo sapiens*. Other allegations include the claims that Neandertals only used locally available material and never traveled beyond 10 kilometers to gather raw materials. Can we claim techniques used by Neandertals in the production of their various artifacts are inferior? The answer to this question would be a resounding no! Human beings occupying various environments usually exploit what is locally available to them. They also adopt techniques for survival that suit their needs, skills, and available resources. The Neandertals survived with their available techniques for many generations, and it can also be observed that their production techniques were never stagnant; they improved with time and graduated from the early Levallois techniques known from the later Acheulian times to refined Levallois technique that culminated in the Châtelperronian tool kit. This development in technology overlapped the Aurignacian industry in some ways. The evidence of adaptability can function to dispel the argument for primitive tradition. Another claim has been that humans traveled longer distances to fetch raw material for their tools—a behavioral modernity—and that this differentiated them from Neandertals. This summation may be read to imply that the longer the distance to the raw material source from the base camp, the more advanced the group. Evidence from many parts of the Old World shows that raw material choices were matters dictated by availability, social ties, and later, trade. There is nothing wrong in using what works for you, especially if it is locally available. Often, human groups with no locally available raw materials resort to long-distance travel or trade for resources only when essential. In some instances, the use of both local raw material and those from far-off places also occurred among Neandertals, as is shown in a study of raw material economics in Apulia, Italy (Spinapolice, 2012)—thus dispelling the notion that only modern humans were capable of long-distance raw material procurement. But it is hard to reason that these make some groups look better or more sophisticated than others. It is fair to argue that evidence for long-distance transportation of raw materials may have evolved due to need, intergroup exchange, which may include marriages, friendships, and so forth. Do not read too much into the need for traveling great distances as a sign of modernity. The weighting on such summations are relative.

Are all Neandertal inventions copycats from modern humans? Enculturation is one of the several possibilities by which human groups adapt new cultural traits that they find useful. This cannot be ruled out in the case of the Neandertals. But it cannot also be argued that it was a one-way street without the burden of definitive problems of the term with the connotation of an imperialistic view. Neighboring groups are more likely to learn from one another than from far-off groups. If we assume that Neandertal social groups were fairly connected with one another, they were likely to learn from one another. Other social groups may also have learned from modern humans (not for researchers who view them as different species); but modern humans also must have learned a thing or two from the Neandertals. It is important to remember that the Neandertals survived the extreme European Ice Ages for many generations before the moderns arrived. Both fossil and genetic evidence support the likelihood of admixture. A scenario that meant more than genetic interchange alone, but also cultural learning between the groups.

Are Neandertals responsible for the Châtelperronian tradition? Likely! The tradition emerged earlier than the Aurignacian tradition associated with modern humans (Caron et al., 2011). It was also fairly widespread and mirrored the early Aurignacian traditions in some ways. If one thinks that there was no species difference between the two groups, the

Châtelperronian ownership becomes less of a problem because it signifies not only a possibility for increased interbreeding (see Zilhão, 2001, 2006), but also enculturation. Because early modern humans may not have perceived that they were entirely different from one another as prescientific Europeans thought, the adoption of traditions and interbreeding may offer us a new way of understanding early humans without the baggage that has cluttered our thoughts for millennia.

THE FATE OF THE NEANDERTALS: THE NEANDERTAL EXTINCTION THEORIES

Neandertal people coexisted with the modern humans for more than 60,000 years before their disappearance. The nature of this disappearance—also termed extinction—has generated debates since the emergence of the group in the fossil record in the early 1800s. There are various competing hypotheses proposed for the extinction of the Neandertals. There is one school of thought that maintains they were phased out by modern humans via competition. This school operates on the premise that modern humans from Africa replaced the Neandertals, with little or sometimes no admixture, depending on whether one is entertaining the Out of Africa or Multiregional hypotheses for the origins of modern humans. Another school also proposes climatic forcing as a significant player in the demise of the Neandertals after surviving for more than 150,000 years. Climate change, accompanied by a lack of proper adaptation on the part of the Neandertals, led to their demise. Another perspective also considers genetic swamping by colonizing moderns who were migrating from out of Africa.

The Neandertal-modern human "competition" hypothesis posits that modern humans had better technology. That the lithic technology had improved, allowing modern humans to access resources more efficiently. For instance, while it is proposed that Neandertals still used short stabbing spears that exposed them to danger, humans were efficient spear throwers, enabling them to kill game from a distance (Churchill et al., 2009). Many members of this school of competition are also proponents of the different-species hypothesis. Therefore, their preoccupation has been with looking for the differences between Neandertals and modern human behavior. The argument on whether Neandertals could be associated with behavioral modernity, especially those exemplified by the Châtelperronian culture, is at the core of this extinction framework. The advent of the Aurignacians (modern humans) is read as an example of behavioral sophistication that put the moderns at an advantage over their Mousterian cousins (e.g., Klein, 2009; Tattersall, 1999). Neandertals in this scheme are believed to be incapable of making symbolic artifacts, items of leisure, and other forms of art, because they missed the ship of the cerebral rubicon (of Clive and Gamble, 1996), a period of speciation of the high-powered brain human from Africa (but see Zilhão et al., 2006 for a contrary opinion).

Another aspect of extinction by competition toys with the possibility of violence via warfare. Marcelline Boule was the first to propose this scenario by stating that modern humans possibly outsmarted the Neandertals due to their being clumsy and dim-witted compared to modern humans represented by the Cro-Magnon people (Boule, 1913, referenced in Tattersall, 1999: 92). The arguments for this possibility have been that humans possessed better tools/weapons that were used to decimate Neandertal populations. The evidence of sharp object trauma on the rib of one of the Neandertals from Shanidar (Shanidar 3) has been used as evidence to drum up support for the warfare hypothesis (e.g., Churchill et al., 2009). As mentioned earlier, this interpretation can go either way: intraspecies violence, or as they say, a case of humans versus Neandertals, but an accident cannot also be ruled out. It is not conclusive. In this school of thought, the Neandertals are seen as an entirely different species, dim-witted, clumsy, and unable to produce fertile hybrids with modern humans; they were outsmarted by better throwing weaponry such as the one that caused the traumatic lesion on the rib of the Shanidar 3 person (Churchill et al., 2009).

Climatic change has also been implicated for contribution to the disappearance of the Neandertals. As palaeoclimatology indicates, the Ice Age began in the Pleistocene some 2.4 mya. The environments in southern Europe was like Africa: it included elephants, hippos, sparse deciduous forests, and the big cats (see Finlayson, 2007). Before the Holocene Epoch, Europe was exposed to thousands of years of an Ice Age environment and climatic fluctuations, with nearly 20 cooling and warming cycles recorded. The Late Pleistocene climate had the warm peaks of the last interglacials (ca. 120,000–125,000 years ago), which curiously coincided with the consensus date estimated for several genetic studies for the origins of modern *Homo sapiens*. The last refinements on language and improvements in technology may have occurred at this time, leading to a substantial cultural advantage in surviving this time of extreme climatic change.

However, an opposite climatic event—that of an extreme drop in world temperatures—occurred around 30,000 years ago. This event is also coincidentally synchronized with the disappearance of the Neandertals. The swings between warm and cold climatic shifts rapidly changed the environment, resulting in the extinction of many plant and animal species that the Neandertals depended on for their survival. Another event triggered by the extreme cooling was the southerly migration of many animals. The migration may have affected northern Neandertal populations that did not migrate. It is also important to note the preference for the warmer southern climes by Neandertals, as evidenced by some of the last members of the group in the Rockshelter of Gibraltar, Spain (Hublin, 1994; Hublin & Trinkaus, 1998), and Lagar Velho, Portugal (Duarte et al., 2009). According to Finlayson and Carrión (2007), climate change— not modern humans—caused the extinction of the Neandertals, especially in the Iberian Peninsula. This event coincides with the coldest period of the Ice Age. A corollary to this hypothesis is an evolutionary explanation that can support the idea that this human group did not adapt well to the changing environment and were naturally selected out. Other workers also implicate the "nonintelligence" and "inferior technology" for cold survival, especially when compared to modern humans, as contributing to the demise of the group. While potentially a plausible hypothesis, climate change alone fails to explain the extinction of Neandertals in the Middle East or eastern Europe.

Another explanation of the disappearance of the Neandertals may be due to "genetic swamping." That is, the distinctive Neandertal type disappeared due to "genetic domination" by the more populous moderns. The Neandertal population may have been low in number due to the climatic restriction of the Ice Age, and strong selection also favored the tropical types. Recent mtDNA data support interbreeding with some one to four percent of Neandertal genes being shared with Europeans and Asians (Green et al., 2010). If present humans show remnants of Neandertal DNA after more than 30,000 years since the last known Neandertal lived, then it is possible that the percent of gene admixture may have been higher at the height of their interaction with modern humans. Evidence of nuclear DNA studies (language gene, pigmentation genes, ABO blood groups, bitter taste gene, microcephaline genes) all point to similarities between Neandertals and modern humans (Krause et al., 2007; Lalueza-Fox et al., 2007, 2008,

2009; Lari, 2010) and that they shared most of the genetic material well before the Ice Age separated them after 400,000 years BP.

The evidence for admixture rules out the question of separate species. Fossil evidence also shows that the Neanderthaloids disappeared in Europe around 30,000 years ago in Zafarraya, Spain (Hublin, 1994; Hublin& Trinkaus, 1998), and possibly as late as 27,000–25,000 years ago in Lagar Velho, Portugal (Duarte et al., 1999), also further supporting the scenario for admixture (Zilhão, 2006). This time period coincided with modern humans represented by the Cro-Magnon in England, also dated to around 30,000 years BP. The Lagar Velho individual has been claimed to show a mixture of both Neandertal and modern human characteristics. The Neandertal traits observed in the four-year-old child from a burial in Lapedo Valley in Portugal include a backward-sloping mandible and a lower limb (tibia) that is proportionally shorter in relation to the femur. The limb proportions show the typical hyperarctic adaptation reminiscent of Neandertals when compared to the modern Europeans, who are similar to Africans in their tropical limb proportions—longer lower limbs. The mandible is also receding in the characteristic Neandertal fashion, but it also shows the presence of a chin. A chin (mental osseum) is one of the anatomical hallmarks of modern humans. The anterior teeth and the thumb phalanges are more modern in the Lagar Velho child. Because of these lines of evidence, it has been suggested that hybridization or interbreeding occurred between the Neandertals and the moderns.

The readings from this range of evidence have implications for explaining the disappearance of the Neandertals. We now know that Neandertals coexisted with modern *Homo sapiens* for nearly 60,000 years. During this time, these fairly disparate populations may have interacted in many ways. The evidence from Lagar Velho offers one example of such interactions—that of interbreeding. If this scenario was common, it is possible to propose that the "extinction" of the Neandertals can be largely biological. How is this scenario possible? We know from biodiversity studies that African or tropical plant and animal species are usually hyperreproductive—that is, they are very fertile. High fertility success implies that these populations are usually higher compared to their temperate counterparts. In the case of Neandertals and modern humans, we can say that the migrating Africans outnumbered the Neandertals, who were probably fewer in number at the time of the glacial maximum 30,000 years ago. The scenario for the disappearance of Neandertals can only be due to the disappearance of the Neandertal phenotype by genetic swamping. This means that because of their lower numbers, selection increasingly favored the modern human phenotype, and this eventually phased out the Neandertal phenotype as we know it. This scenario also supports the presence of several shared genetic traits with Neandertals, making them part of the same species.

CONCLUSIONS

Currently, the paleoanthropological views about Neandertals are changing considerably in the face of new data and a more balanced interpretation of the evidence. This will help give humanity back to the Neandertals. Especially significant is the new evidence that we share most of the genetic materials for pigments, language, bitter taste, and the ABO blood groups. All this genetic evidence points to a shared ancestry before the split and evolution of the Neandertal morphological phenotype. If these new data hold true, it can be pointed out that the Neandertal phenotype was one that was a response to isolation due to Ice Age climatic forcing of responses to Bergmann's and Allen's rules. Even total genome analyses (Green et al., 2010) show some evidence of shared genetic material in non-Africans. As mentioned earlier, present humans show polymorphisms in their genetic makeup. The presence of sickle-cell trait, lactase persistence, and uneven distribution of ABO blood groups in human populations may be important in helping to keep the presence of some form of "unique" traits of microcephalin genes among Neandertals and even variants of the MC1R gene. The slight difference in these genetic materials in the Neandertal—and by extension, human population—may now help normalize such cases and not reading too much variation (different species) between the human populations and those with the Neandertal morph.

Another line of evidence supporting their humanity comes from the several skeletal lesions that commonly afflicted the Neandertals. These pathologies have been interpreted to represent a compassionate group of people who cared for their sick and the old. It is only through care that the ailing members of the group could last long enough for the lesions to leave remodeled bone remains for us to study. The survival of the Old Man from La Chappelle, who lived to be over 55 years of age with a debilitating paralysis, is a compelling example. Other evidence indicates that the Neandertals' longevity profiles were not any different from the rest of the prehistoric and even living hunter-gatherer groups. Many of the deaths at adult or old adult ages also imply that many members probably died at their reproductive ages, but since reproductive activities in hunter-gatherer lifestyle start early, they were likely successful because they produced enough offspring to help keep their population numbers stable, at least until after 30,000 years, when more tropical people arrived in their turf.

Many researchers now maintain that evidence for temperate adaptation may only be seen in the stature, especially the distal lower limbs and perhaps the torso structure and muscle markings representing a bulky frame. Other evidence such as the large paranasal aperture and other pneumatized structures such as the frontal, ethmoidal, and maxillary sinuses were not due to cold adaptation, but very tropical; that their large sizes in Neandertals were due to scaling. Evidence from other animals, including mice, indicates that exposure to cold actually leads to a reduction in these structures. The presence of these enlarged structures therefore are now considered retentions of shared characteristics with more tropical people. This further reduced the number of traits that have been used to separate the morphological and perhaps behavioral distance between Neandertals and modern humans.

The debates on the language abilities of the Neandertals are close to being settled. While earlier studies focused on the hyoid bone and its position in the throat, new evidence from the FOXP2 gene now supports the case for similarity between the Neandertals and us. This gene is known to be responsible for some linguistic functions and the development of some of the facial muscles responsible for speech production. It may have functioned in the same way in Neandertals. Besides, it has been pointed out that earlier studies which placed the Neandertal hyoid higher in the throat, therefore making it impossible for them to produce all the ranges of human sounds, were erroneous. It compared an adult Neandertal specimen from La Chappelle with a human neonate and a chimp, not an adult human. Just why this study was conducted in this way is puzzling and illogical; it ignored the fact that the structural positioning of the hyoid and the larynx lower in the throat was due to bipedal influence, as well as developmental factors. Treating a bipedal being with a non-bipedal neonate and chimpanzee was an oversight at the time. The morphology of the Neandertal hyoid is also similar to that of humans. These two lines of evidence now point toward a solid case for speech ability by the Neandertals.

The behavioral repertoire of the Neandertals is expansive. The most significant aspect is their burial practice. It is now acknowledged that Neandertals intentionally buried their dead long before there was evidence of similar practice among anatomically modern humans. Their graves included goods of various categories and in some cases, flowers. Burial practice often is associated with belief in an afterlife or some kind of religion. Burials also require planning; therefore, a significant indicator of forethought. Forethought, belief in an afterlife, burials, and compassion for the departed members are hallmarks of humanity—making a strong case for the human nature of Neandertals.

Neandertals also survived the harsh climes of Ice Age Europe using what we consider today as primitive technology. This feat is commendable! It led the Neandertals to improve their shelter, use fire regularly, and even make clothes to keep them warm. These technological innovations indicate that they were adaptable and did not have a stagnant cultural know-how, as some archaeologists have claimed in regard to their Mousterian lithic industries. Some evidence points out that the Mousterian tradition persisted in the "same" fashion for long periods with little change. Others archaeologists also see that there were diachronic changes in the technology, and some differences were also regional, reflecting specific cultural styles and adaptations. At the core of these technological debates is the period marking the end of the Middle Paleolithic and the transition to the Upper Paleolithic. Some claims are that the main industries of the early Upper Paleolithic, the Châtelperronian, and the Aurignacian were distinct enough, and the association of the Neandertals with the Châtelperronian traditions has been contested. The claim is that the tradition mimics the Aurignacian, despite the fossil evidence associated with the Châtelperronian indicating undisputable Neandertal fossils. The beliefs that these are intrusions of the Upper Aurignacian levels are common. Other alternative explanations for the presence of Châtelperronian in association have been claimed to be due to acculturation by modern humans. Recent revisions on the interpretation of the two cultural levels indicate that in Arcy, for instance, the Châtelperronian levels that are lower (level X) show more artifacts of that tradition, and there is a lull, or reduction, in the artifact counts in the two succeeding Châtelperronian horizons (IX and VIII). Even the lower Aurignacian levels at Arcy (VII) do not show the presence of artifact concentration as at the lower Châtelperronian level. This evidence has led some workers to assert that Neandertals are responsible for the Châtelperronian innovations. If this position is maintained, it shows further that Neandertals were "progressive" in their industrial techniques—including new raw materials such as bones and ivory, as well as making items of leisure and trade like pendants with teeth, bone, ivory, or shell materials. If the presence of these items is the hallmark of humanity, then it is safe to say that the Neandertal morphotype achieved human features before their morph "melted" into anatomical moderns. As physical anthropologist Dave Phillips stated: "Neanderthals were human. They buried their dead, used tools, had a complex social structure, employed language, and played musical instruments. Neanderthal anatomy differences are extremely minor and can be for the most part explained as a result of a genetically isolated people that lived a rigorous life in a harsh, cold climate (Phillips, 2000, "Neanderthals Are Still Human")."

BIOGRAPHY

Dr. Nasser Malit earned his Ph.D. from the State University of New York (SUNY) at Binghamton in 2009. His research interests involve the evolution of humans in the Quaternary. His current research interests focus on the study and evolution of *Homo erectus* where he employs morphometric methods in understanding the biology and variability among these defunct hominins. He is also committed to the reevaluation of the Middle Pleistocene hominin record from Africa. This aspect of research has driven his interest in the European fossil record especially in the evolution of the Neandertals. He believes that because the Neandertals have yielded more fossils, DNA and behavioral evidence than any other hominin group, they provide a good window into understanding our origins. He presently teaches biological anthropology courses at the Department of Anthropology, SUNY Potsdam.

REFERENCES

The History and Chronology of Neandertal Finds

Klein, Richard (2009). *The Human Career: Human Biological and Cultural Origins.* Third Edition. University of Chicago Press.

Schrenk, Friedemann, & Müller, Stephanie (2009). *The Neanderthals: Peoples of the Ancient World.* Routledge.

Speth, J. D. (2004). News flash: Negative evidence convicts Neandertals of gross mental incompetence. *World Arch.* 36 (4): 519–526.

Tattersall, Ian (1999). *The Last Neanderthal: The Rise, Success, and Mysterious Extinction of Our Closest Human Relatives.* Revised Edition. American Museum of Natural History, West View Press.

Trinkaus, E., & Shipman, P. (1993). *The Neandertals: Changing the Image of Mankind.* New York: Knopf.

Phylogenetics and Origins Debates

Dean, D., Hublin, J. J., Holloway, R., & Ziegler, R. (1998). On the phylogenetic position of the pre-Neandertal specimens from Reilingen, Germany. *J. Hum. Evol.* 34, 485–508.

Hublin, J. J. (2009). The origins of Neandertals. *PNAS,* 106 (38): 16022–16027.

Smith, F. H. (1991) The Neandertals: Evolutionary dead ends or ancestors of modern humans? *J. Anth. Research.* 47: 219–238.

Neandertal Morphology

Ackermann, R. R., & Cheverud, J. M. (2004). Detecting genetic drift versus selection in human evolution. *Proc. Natl. Acad. Sci. USA* 101: 17946–17951.

Antón, S. C. (1994). Mechanical and other perspectives on Neanderthal craniofacial morphology. In R. S. Corruccini & R. L. Ciochon (Eds.), *Integrative Paths to the Past* (pp. 677–695). Englewood Cliffs, NJ: Prentice Hall.

Arensburg, B., Schepartz, L. A., Tillier, A. M., Vandermeersch, H., Duday, H., & Rak, Y. (1990). A reappraisal of the anatomical basis for speech in Middle Palaeolithic hominids. *Am. J. Phys. Anthropol.* 83: 137–46.

Arensburg, B., Tillier, A. M., Vandermeersch, B., Duday, H., Schepartz, L. A., & Rak, Y. (1989). A Middle Paleolithic human hyoid bone. *Nature* 338: 758–760.

Bar-Yosef, O., Vandermeersch, B., Arensburg, B., Belfer-Cohen, A., Goldberg, P., Laville, H., Meignen, L., Rak, Y., Speth, J. D., Tchernov, E., Tillier, A. M., & Weiner, S. (1992). The excavations in Kebara Cave, Mt. Carmel. *Curr. Anthropol.* 33: 497–550.

Bailey, S. E., & Lynch, J. M. (2005). Diagnostic Differences in Mandibular P4 Shape between Neandertals and Anatomically Modern Humans. *Am. J. Phys. Anthropol.* 126: 268–277.

Berger, T., & Trinkaus, E. (1995) Patterns of trauma among the Neandertals. *J. Archaeol. Sci.* 22: 841–852.

Cowgill, L. W., Trinkaus, E., & Zeder, M. A. (2007) Shanidar 10: A Middle Paleolithic immature distal lower limb from Shanidar Cave, Iraqi Kurdistan. *J. Hum. Evol.* 53: 213–223.

Dawson, J. E., & Trinkaus, E. (1997). Vertebral Osteoarthritis of the La Chappelle-aux-Saints 1 Neanderthal. *J. Arch. Sci.* 24: 1015–1021.

Deacon, T. (1994). The Human Brain. In: Jones, S., R. Martin, & D. Pilbeam (eds.), *The Cambridge Encyclopedia of Human Evolution.* Cambridge University Press.

DeSilva, J. M. (2011). A shift toward birthing relatively large infants early in human evolution. *PNAS*, 108 (3): 1022–1027.

Endicott, P., Ho, S. Y. W., & Stringer, C. (2010). Using genetic evidence to evaluate four palaeoanthropological hypotheses for the timing of Neanderthal and modern human origins. *J. Hum. Evol.* 59: 87–95.

Falk, D. (1975). Comparative Anatomy of the Larynx in Man and the Chimpanzee: Implications for Language in Neandertal. *Am. J. Phys. Anthropol.* 43: 123–132.

Franciscus, R. G. (2009). When did the modern human pattern of childbirth arise? New insight from an old Neandertal pelvis. *PNAS*, 106 (23): 9125–9126.

Frayer, D. (1993). On Neanderthal Crania and Speech: "Response to Lieberman." *Current Anthropol.* 34: 721.

Golovanova, L. V., Hoffecker, J. F., Kharitonov, V. M., & Romanova, G. P. (1999). Mezmaiskaya Cave: A Neanderthal occupation in the Northern Caucasus. *Curr. Anthropol.* 40: 77–86.

Gómez-Olivencia, A., Eaves-Johnson, K. L., Franciscus, R. G., Carretero, J. M., & Arsuaga, J. L. (2009). Kebara 2: New insights regarding the most complete Neandertal thorax. *J. Hum. Evol.* 57: 75–90.

Holliday, T. (1997). Postcranial evidence of cold adaptation in European Neanderthals. *Am. J. Phys. Anthropol.* 104: 245–258.

Holton, N. E., & Franciscus, R. G. (2008). The paradox of a wide nasal aperture in cold-adapted Neandertals: A causal assessment. *J. Hum. Evol.* 55, 942–951.

Hublin, J. J. (1998) In *Neandertals and Modern Humans in Western Asia*, eds. Akazawa T., Aoki, K., Bar-Yosef, O. (Plenum, New York), pp. 295–310.

Hublin, J. J., Barroso, Ruiz C., Medina, Lara P., Fontugne, M., Reyss, J.-L. (1995) The Mousterian site of Zafarraya (Andalucia, Spain): Dating and implications on the Palaeolithic peopling processes of Western Europe. *Comptes Rendus de l'Academie des Sciences Paris* 321: 931–937.

Klein, Richard (2009). *The Human Career: Human Biological and Cultural Origins.* Third Edition. University of Chicago Press.

Lieberman, P. (1993). On the Kebara KMH 2 Hyoid and Neanderthal Speech. *Curr. Anthropol.* 34: 172–175.

Lieberman, P., & Crelin, E. S. (1971). On the speech of Neandertal man. *Linguistic Inquiry*, 2: 203–222.

Marlowe, F. (2005). Hunter-gatherers and human evolution. *Evol. Anthrop.* 14: 54–67.

Odwak, H. (2000). Long bone robusticity and claviculohumeral proportions of the Amud 7 Neandertal baby. *Am. J. Phys. Anthropol.* 30 (Suppl): 241.

Plummer, T., & Bishop, L. (1994). Hominid paleoecology at Olduvai Gorge, Tanzania, as indicated by antelope remains. *J. Hum. Evol.* 27, 47–75.

Ponce de Léon, M. S., Golovanova, L., Doronichev, V., Romanova, G., Akazawa, T., Kondo, O., Ishida, H., Zollikofer, C. P. E. (2008). Neandertal brain size at birth provides insights into the evolution of human life history. *PNAS*, 105 (37): 13764–13768.

Rae, T. C., Koppe, T., & Stringer, C. B. (2011). The Neanderthal face is not cold adapted. *J. Hum. Evol.* 60(2): 234–239.

Rak, Y., & Arensburg, B. (1987). Kebara 2 Neandertal pelvis: First look at a complete inlet. *Am. J. Phys. Anthropol.* 73: 227–231.

Rosas, A. (2001). Occurrence of Neandertal Features in Mandibles from the Atapuerca-SH Site. *Am. J. Phys. Anthropol.* 114, 74–91.

Ruff, C. B. (2000). *Biomechanical Analyses of Archaeological Human Skeletons*. In: *Biological Anthropology of the Human Skeleton*. Katzenberg, A., & Saunders, S. R., eds., pp. 71–102. New York: Alan R. Liss.

Serrat, M. A., King, D., & Lovejoy, C. O. (2008). Temperature regulates limb length in homeotherms by directly modulating cartilage growth. *PNAS*, 105 (49): 19348–19353.

Smith, F. (1978). Evolutionary Significance of the Mandibular Foramen Area in Neandertals. *Am. J. Phys. Anthropol.* 48, 523–532.

Stringer, C., & Gamble, C. (1993). *In Search of the Neanderthals*. Thames and Hudson.

Tattersall, Ian (1999). *The Last Neanderthal: The Rise, Success, and Mysterious Extinction of Our Closest Human Relatives*. Revised Edition. American Museum of Natural History, West View Press.

Tappen, N. C. (1985). The dentition of the old man of La Chappelle-aux-Saints and inferences concerning Neandertal behavior. *Am. J. Phys. Anthropol.* 67 (1): 43–50.

Trinkaus, E. (2012). Neandertals, early modern humans, and rodeo riders. *J. Arch. Sci.* 39 (12): 3691–3693.

Trinkaus, E. (2011). The postcranial dimensions of the La Chappelle-aux-Saints 1 Neandertal. *Am. J. Phys. Anthropol.* 145(3): 461–468.

Trinkaus, E. (1995a). Neandertal mortality patterns. *J. Arch. Sci.* 22: 121–142.

Trinkaus, E., & Shipman P. (1992). *The Neanderthals: Changing the Images of Mankind*. Alfred A. Knopf, New York.

Trinkaus, E. (1985). Pathology and the posture of the La Chappelle-aux-Saints Neandertal. *Am. J. Phys. Anthropol.* 67 (1):19–41.

Trinkaus, E. (1984). Neandertal pubic morphology and gestation length. *Curr. Anthropol.* 25: 509–514.

Trinkaus, E. (1983). Neandertal postcrania and the adaptive shift to modern humans. In: Trinkaus, E. *The Mousterian Legacy: Human Biocultural Change in the Upper Pleistocene*, pp. 165–200.

Trinkaus, E. (1980). Sexual differences in Neanderthal limb bones. *J. Hum. Evol.* 9, 317–397.

Trinkaus, E. (1978). Hard times among the Neanderthals. *Natural History* 87: 58–63.

Vinicius, L. (2005). Human encephalization and developmental timing. *J. Hum. Evol.* 49: 762–776.

Weaver, T. D. (2009). The meaning of Neandertal skeletal morphology. *PNAS*, 106 (38): 16028–16033.

Weaver, T. D., & Hublin, J. J. (2009). Neandertal birth canal shape and the evolution of human childbirth. *Proc. Natl. Acad. Sci. USA* 106: 8151–8156.

Neandertal Genetics

Enard, W., Przeworski, M., Fisher, S. E., Lai, C. S., Wiebe, V., Kitano, T., Monaco, A. P., & Pääbo, S. (2002). Molecular evolution of FOXP2, a gene involved in speech and language. *Nature* 418, 869–872.

Gluckman, P., Beedle, A., & Hanson, M. (2009). *Principles of Evolutionary Medicine*. Oxford University Press, USA.

Green, R. E., Krause, J., Briggs, A. W., et al. (53 coauthors). 2010. A draft sequence of the Neandertal genome. *Science* 328: 710–722.

Krause, J., Lalueza-Fox, C., Orlando, L., Enard, W., Green, R. E., Burbano, H. A., Hublin, J. J., Bertranpetit, J., Hänni, C., de la Rasilla, M., et al. (2007). The derived FOXP2 variant of modern humans was shared with Neanderthals. *Curr. Biol.* 17, 1908–1912.

Lalueza-Fox, C., Rompler, H., Caramelli, D., Staubert, C., Catalano, G., Hughes, D., Rohland, N., Pilli, E., Longo, L., Condemi, S., et al. (2007). A melanocortin 1 receptor allele suggests varying pigmentation among Neanderthals. *Science* 318, 1453–1455.

Lalueza-Fox, C., Gigli, E., de la Rasilla, M., Fortea, J., Rosas, A., Bertranpetit, J., & Krause, J. (2008). Neanderthal paleogenomics in the ABO blood group gene. *BMC Evol. Biol.* 8, 342.

Lalueza-Fox, C., Gigli, E., de la Rasilla, M., Fortea, J., and Rosas, A. (2009). Bitter taste perception in Neanderthals through the analysis of TAS2R38 gene. *Biol. Lett.* 5, 809–811.

Lai, C. S., Fisher, S. E., Hurst, J. A., Vargha-Khadem, F., and Monaco, A. P. (2001). A forkhead-domain gene is mutated in a severe speech and language disorder. *Nature* 413, 519–523.

Lari, M., Rizzi, E., Milani, L., Corti, G., Balsamo, C., et al. (2010) The *Microcephalin* Ancestral Allele in a Neanderthal Individual. *PLoS ONE* 5(5): e10648. doi:10.1371/journal.pone.0010648.

Trinkaus, E. (2007). Human Evolution: Neandertal Gene Speaks out. *Curr. Biol.* 17, R917–R919.

Vargha-Khadem, F., Gadian, D. G., Copp, A., & Mishkin, M. (2005). FOXP2 and the neuroanatomy of speech and language. *Nat. Rev. Neurosci.* 6, 131–138.

Neandertal Behavior and the Archaeological Record

Caron, F., D'Errico, F., Del Moral, P., Santos, F., & Zilhão, J. (2011) The Reality of Neandertal Symbolic Behavior at the Grotte du Renne, Arcy-sur-Cure, France. *PLoS ONE* 6(6): e21545.

d'Errico, F., Zilhão, J., Baffier, D., Julien, M., & Pelegrin, J. (1998). Neanderthal Acculturation in Western Europe? A Critical Review of the Evidence and Its Interpretation. *Curr. Anthropol.* 39, pp. S1–S44.

d'Errico, F., Henshilwood, C., Vanhaeren, M., & van Niekerke, K. 2005. Nassarius kraussianus shell beads from Blombos Cave: Evidence for symbolic behaviour in the Middle Stone Age. *J. Hum. Evol.* 48(1): 3–24.

Duarte, C., Maurício, J., Pettit, P., Souto, P., Trinkaus, E., van der Plicht, H., & Zilhão, J. 1999: The early Upper Paleolithic human skeleton from the Abrigo do Lagar Velho (Portugal) and modern human emergence in Iberia, *PNAS*, 1999 Jun 22; 96(13): 7604–7609.

Gamble, C. (1993). *Timewalkers: The Prehistory of Global Colonization*. Penguin Books, USA.

Green, R. E., Krause, J., & Briggs, A. W. (2010). A draft sequence of the Neandertal genome. *Science* 328, 710–722.

Gowlett, J. 1994. Early human mental abilities. In: Jones, S., R. Martin, and D. Pilbeam (eds.). *Ancestors: The Hard Evidence*. New York: Alan R. Liss Inc.

Henshilwood, C. S., d'Errico, F., Yates, R., Jacobs, Z., Tribolo, C., Duller, G. A., Mercier, N., Sealy, J. C., Valladas, H., Watts, I., & Wintle, A. G. (2002). Emergence of modern human behavior: Middle Stone Age engravings from South Africa. *Science* 295 (5558): 1278–1280.

Higham, T., Jacobi, R., Julien, M., David, F., Basell, L., et al. (2010). Chronology of the Grotte du Renne (France) and implications for the context of ornaments and human remains within the Châtelperronian. *PNAS*107: 20234–20239.

Hublin, J. J., Spoor, F., Braun, M., Zonneveld, F., & Condemi, S. (1996). A late Neanderthal associated with Upper Palaeolithic artefacts. *Nature* 381, 224–226.

Klein, R. (1999). *The Human Career*. University of Chicago Press.

Lewin, R. (1998). *The Origin of Modern Humans*. Scientific American Library.

McBrearty, S., & Brooks, A. S. (2000). The revolution that wasn't: A new interpretation of the origin of modern human behavior. *J. Hum. Evol.* 39, 453–563.

Mourre, V., Villa, P., & Henshilwood, C. 2010. Early use of pressure flaking on lithic artifacts at Blombos Cave, South Africa. *Science* 330: 659–662.

Pettitt, P. B. (2002). The Neanderthal dead: Exploring mortuary variability in Middle Palaeolithic Eurasia. *Before Farming* 1 (4): 1–26.

Semaw, S., Renne, P., Harris, J. W. K., Feibel, C. S., Bernor, R. L., Fesseha, N., & Mowbray, K. (1997). 2.5-million-year-old stone tools from Gona, Ethiopia, *Nature* 385, 333–336.

Spinapolice, E. E. (2012). Raw material economy in Salento (Apulia, Italy): New perspectives on Neanderthal mobility patterns. *J. Arch. Sci.* 39, 680–689.

Yellen, J. E., Brooks, A. S., Cornelissen, E., Mehlman, M. J., & Stewart, K. (1995). "A Middle Stone Age worked bone industry from Katanda, Upper Semliki Valley, Zaire." *Science* 268 (5210): 553–556.

Zilhão, J. 2001: The Lagar Velho Child and the Fate of the Neanderthals, *Athena Review*, vol. 2, no. 4, 33–39.

Fate of the Neandertals

Churchill, S. E., Franciscus, R. G., McKean-Peraza, H. A., Daniel, J. A., & Warren, B. R. (2009). Shanidar 3 Neandertal rib puncture wound and paleolithic weaponry. *J. Hum. Evol.* 57: 163–178.

Duarte, C., Maurício, J., Pettitt, P. B., Souto, P., Trinkaus, E., van der Plicht, H., & Zilhão, J. (1999). The early Upper Paleolithic human skeleton from the Abrigo do Lagar Velho (Portugal) and modern human emergence in Iberia. *Proc. Natl. Acad. Sci.* 96: 7604–7609.

Finlayson, C., & Carrión, J. S. (2007). Rapid ecological turnover and its impact on Neanderthal and other human populations. *Trends in Ecology & Evolution* 22, 213–222.

Hublin, J. J., & Trinkaus, E. (1998). The Mousterian human remains from Zafarraya (Granada, Spain). *Am. J. Phys. Anthropol.* suppl. 26: 122–123.

Hublin, J. J. (1994). The late Neandertal site of Zafarraya. AEQUA *Monografias* 2: 61–70.

Zilhão, J. (2006). Neandertals and moderns mixed, and it matters. *Evol. Anthropol.* 15: 183–195.

Zilhão, J., d'Errico, F., Bordes, J.-G., Lenoble, A., Texier, J.-P., et al. (2006). Analysis of Aurignacian interstratification at the Châtelperronian-type site and implications for the behavioral modernity of Neandertals. *PNAS*, 103: 12643–12648.

Section Three

ETHICS AND CAREERS: HUMAN ISSUES OF TODAY

ETHICAL CONSIDERATIONS

By Barbara J. Little

Anyone who isn't confused here doesn't really understand what's going on.
—Anonymous

Archaeological ethics can be complex and confusing. Responsibilities extend to the archaeological record, many different publics, and one's colleagues. As Randall McGuire points out, the development of cultural resource management over the last few decades has turned archaeology into a business, something that it really hadn't been before. Such growth raises ethical issues involved with "the relationship of research to legal compliance, the relationship of academic archaeology to business, the training of students, publication, and public outreach" (McGuire 2003: viii). In the academic realm, ethical issues surround teaching and treatment of students. Similar issues extend to one's coworkers. Archaeologists have slowly become aware of serious imbalances and stereotypes (based on race and sex especially) that adversely impact the whole field by affecting the kind of work individuals are trained and encouraged to do.

In this chapter I provide very brief characterizations of some categories of ethical issues. At the end of this chapter is a list of major professional organizations with ethics statements and URLs to find their current statements on-line.

CONSERVATION OF THE ARCHAEOLOGICAL RECORD

Even though the archaeological record in a general sense is being continually created, archaeological remains are nonrenewable in the sense that such records of the past cannot be re-created. Many sites are destroyed or damaged by economic development—ranging from agriculture, to dam and road construction, to suburban subdivisions and urban renewal—or by looters who mine sites for artifacts they can sell or keep for their own collections. Casual collectors, recreational off-road enthusiasts, or divers can also damage sites, often unwittingly.

Some governments respond to the looting problem with creative and sometimes effective solutions, but the motivation for collecting ensures that the problem will continue. In the United States, the National Park Service and the Bureau of Land Management, for example, run sting operations to catch traffickers in illegally acquired artifacts. Vigorously prosecuting violations of the law is one tactic, but of course most archaeological remains are not protected by law. Public outreach and encouragement of public stewardship is another tactic for preservation. Through training programs for amateur archaeologists, several states accomplish a great deal of professional-quality work and dramatically expand public awareness of and concern for the looting problem. Another tactic is to remove sites from the marketplace to reduce the likelihood of looting. Based on the principle of the American sanctity of private property, the Archaeological Conservancy as a private foundation protects sites by purchasing them.

Site preservation was not always a widespread concern in the profession. However, rapidly increasing threats to sites caught the attention of both the profession and the public. Archaeologists have come to recognize their responsibilities to the record of human heritage and, importantly, to the peoples whose heritage is chronicled in archaeological sites. That responsibility extends to collections made from sites, as well as to the sites themselves.

Archaeology as cultural resource management (CRM) done under public law does not often result in the preservation of sites in place. Instead many sites are identified and judged insignificant according to legal standards and professional judgment. If judged insignificant, then sites are destroyed to make way for roads, pipelines, or other public projects. Alternatively, sites may be judged significant and be excavated in order to recover important data. Rarely are sites preserved in place if they are in the way of a development project. The stakes are high. Practicing archaeologists have an important ethical responsibility to be well informed and to do their work conscientiously. Their recommendations reverberate well into the future (see especially the ethics statements of the American Cultural Resources Association).

Because much archaeological fieldwork in the U.S. today is undertaken to comply with public laws, archaeologists are accountable for the results of their work, often paid for by public funds. If the results of the work are not taken care of properly, it is difficult to justify spending public money. Archaeological curation has become a crisis, due to widespread and long-standing lack of attention by archaeologists to their ethical responsibilities for the long-term care and management of archaeological collections. Collections include not only artifacts, but also noncultural materials such as soil samples, radiocarbon samples, and floral and faunal material; associated records such as field notes, maps, photos, and laboratory data; digital data; and research results such as site reports and publications. Everyone involved in creating collections is responsible for their proper curation. Without proper care, the essential archaeological information is likely to be lost because collections will disintegrate, disappear, or become so disorganized that they are unusable.

DIVERSE PUBLICS

Archaeological places and the collections that come from their study are not simply resources full of data, but are places and objects of cultural heritage to living people. Using sites and artifacts for such purposes as education, community cohesion, entertainment, and economic development can lead to widespread benefits but also to conflict. For example, sites can be places of economic potential as heritage tourism destinations. Resulting interest in development can conflict with preservation or the desires of other interested publics, some of whom might not want visitors.

The use and abuse of science is a long and disturbing story. Scientific racism has left a bitter legacy, including the treatment of Native American human remains, which has been a major source of conflict between archaeologists and Native Americans. The ongoing tension between archaeology and the people whose past is studied concerns the ownership of the past and the worldwide issue of intellectual property rights. It is part of the ethical code of most professional archaeologists that the past belongs to everyone, but ethics and values often differ between tribes and professional archaeologists.

Historical archaeologists are not as directly affected by the Native American Graves Protection and Repatriation Act (NAGPRA) as their colleagues who study earlier Native American cultures. However, the ethics of dealing with descendant communities and the interested public apply to all archaeologists, regardless of the legal framework. The African Burial Ground project in New York City—the source of my *sankofa* theme—has been important for historical archaeologists because it brought to the forefront for discussion and debate our responsibilities to descendant communities.

PUBLIC OUTREACH

Professional archaeologists often criticize each other for their poor record in communicating with the general public. Because public outreach has not been part of professional academic responsibilities, public attitudes toward collecting antiquities and ignorance about the real cost of looting have continued to counter efforts at saving sites.

Anyone reading the literature on public archaeology will find that much of archaeologists' interest in public outreach stems from the need to protect and preserve archaeological resources. It should not be surprising, then, that the benefits of preserving sites often are couched in terms of benefit to archaeology through the creation of a public interested in and supportive of archaeology. In some circles, it has become fashionable for archaeologists to criticize the discipline and each other for their supposed limited self-interest in fighting for the protection of archaeological sites. But looting and destroying archaeological sites and other cultural remains is troublesome, and not only to archaeologists. We lose much more than a future for archaeologists with every destroyed site, every record of human achievement.

In the case of artifact trafficking, there is clearly a collision of ethical systems at work. Archaeologists might define it as a clash between business for profit and knowledge for the public good. Or perhaps it is a clash between aristocratic values of privileged personal ownership and democratic notions of common heritage. Where does the entitlement come from that allows one segment of the population to buy and thereby destroy another's history or sense of nation? Do we have (or want) common ground, or is world history just another commodity?

Preserving sites is essential if we are to preserve the public benefits of archaeology as national and local heritage. The malicious destruction of cultural identity came to worldwide attention (again) with the purposeful destruction of the Bamiyan Buddhas in Afghanistan in 2001. The Taliban government was very explicit about its plans to destroy another religion's icons and cultural heritage when they announced their plans and then proceeded to destroy these monumental sculptures carved in the third and fifth centuries.

Archaeologists and curators have a responsibility to be aware of the ethical dilemmas that accompany armed conflict and its aftermath. Some of the unanticipated and unplanned-for consequences of the United States' wars in Afghanistan and Iraq has been the looting of museums, destruction and looting of archaeological sites, and the selling of artifacts on the international market. The Hague Convention of 1954 resulted from long-standing international efforts to protect cultural heritage in times of war. Its preamble states: "Damage to cultural property belonging to any people whatsoever means damage to the cultural heritage of all mankind, since each people makes its contribution to the culture of the world."

There is a direct connection between the destruction of sites by looters and the willingness of museums and private collectors to pay for the looted artifacts. Underlying tensions between the goals of museums and archaeology and their practitioners flare up around this issue, but the ethics of collecting need to be confronted by both serious and casual collectors as well.

There is more to public outreach than convincingly arguing for the preservation of sites and collections. The meaning of public archaeology is changing rapidly as archaeologists work more directly and more cooperatively with communities. Consider as well the challenges of making decisions about how to treat historic properties other than archaeological sites. Historic preservation is not just about preserving beautiful old buildings. It is also about the social responsibility of recognizing, preserving, and interpreting the full and inclusive breadth of our history. Archaeology's long-standing role in the broader world of historic preservation connects archaeologists to those decisions. Clearly, not every archaeological site or every standing structure can be preserved. The choices that are made daily by archaeologists and their colleagues in history and architecture require careful ethical consideration.

In their introduction to *Ethical Issues in Archaeology,* Larry Zimmerman, Karen Vitelli, and Julie Hollowell-Zimmer remind us that codes of ethics are neither static nor necessarily complete or final. Ethical practice needs to be based on ongoing critical reflection, discussion, and debate, and on an evolving process for reaching resolutions. I include URLs for various organizations' codes of conduct here rather than in the Further Readings section because I want you to consider them as part of this chapter. Ethics evolve to meet new challenges and therefore the organizations' websites will give you their most up-to-date versions. As you read them, you will probably notice that there are some conflicts between different organizations' statements. Some privilege the desires of associated groups of people; others privilege the archaeological record and associated information. Grappling with ethical issues can be frustrating. It is difficult but essential for anyone affected by these issues (that would be most of us) to work to apply ethical principles to the specific circumstances with which we are confronted.

ETHICAL CODES OF PROFESSIONAL ORGANIZATIONS

United States

American Anthropological Association Code of Ethics http://www.aaanet.org/committees/ethics/ethcode.htm

American Association of Museums Code of Ethics for Museums http://www.aam-us.org/museumresources/ethics/coe.cfm

American Cultural Resources Association Code of Ethics http://www.acra-crm.org/Ethics.html

Archaeological Institute of America Code of Ethics http://www.archaeological.org/pdfs/AIA_Code_of_EthicsA5S.pdf

National Council on Public History Ethics Guidelines http://www.ncph.org/ethics.html

Register of Professional Archaeologists' Code of Conduct http://www.rpanet.org/

Society for American Archaeology Principles of Archaeological Ethics http://www.saa.org/ABOUTSAA/COMMITTEES/ethics/principles.html

Society for Applied Anthropology Ethical and Professional Responsibilities http://www.sfaa.net/sfaaethic.html

Society for Historical Archaeology Ethical Principles http://www.sha.org/About/ethics.htm

International

Australian Association of Consulting Archaeologists Code of Ethics http://www.aacai.com.au/codeofethics.html

Canadian Archaeological Association Principles of Ethical Conduct http://www.canadian-archaeology.com/conduct.lasso

European Association of Archaeologists Code of Practice and Principles of Conduct http://www.e-a-a.org/eaacodes.htm

International Council of Museums Code of Ethics for Museums http://icom.museum/ethics.html

UNESCO International Code of Ethics for Dealers in Cultural Property http://portal.unesco.org/culture/en/ev.php-URL_ID=13095&URL_DO=DO_TOPIC&URL_SECTION=201.html

World Archaeological Congress First Code of Ethics http://www.worldarchaeologicalcongress.org/site/about_ethi.php

ARCHAEOLOGY

Introductory Guide for Classroom and Field

By Ellis E. McDowell-Loudan

As a starting point for a book that introduces readers to a topic, we shall begin with a series of questions. First, what is it that you like about archaeology and why did you decide to study it? Is it because archaeology consists of sensational and remarkable discoveries in far away places? Do you envision studying remains of early settlers in your region? Perhaps you found bits of old crockery or odd pieces of corroded metal in your yard that do not resemble anything you use today, and were excited about what they imply about previous residents.

Has there been a controversial new shopping mall built in your home town, which caused the demolition of a popular playground or an intriguing old house? Were archaeologists called in to carry out a study before the mall was built? Did they find some amazing clues to ways the land had been used long before your town developed? Sometimes when this happens, the archaeologists involved provide programs for the local historical society. Did you happen to attend one of these and get inspired to learn more about archaeology?

Once there is an exciting archaeological find, there may be a stream of telephone calls and letters arriving at the newspaper, television, historical society, or local college inquiring about updates or offering similar types of information. Sometimes this leads to further discoveries. Occasionally, as a result of this type of publicity, new archaeological sites or collections of artifacts are reported to someone who was involved in the archaeological project. Perhaps you heard about some of these types of discoveries and their aftermath, and want to be part of the exploration and future excitement!

Usually it is the mysterious and unusual find and not the commonplace discoveries that spur interest and publicity. These are the events that we hear the most about, and in some areas, these reports are followed by other reports from other areas because the newspapers or other media recognize public interest in them. When a number of these events are reported in quick succession, readers and viewers may get the impression that there is a steady stream of captivating discoveries, all the time. It is this expectation of constant discovery that draws many people to archaeology.

Truly, there are pieces of our heritage all around us. However, many times these clues to the past are scattered, disrupted by more recent activities, or viewed as too commonplace for much attention to be paid to them. Much depends upon the area, the current interests of people in those regions, the potential for their destruction, and the prospects for funding study of them.

Local, state, and national centennial or bicentennial celebrations are examples of events that tend to trigger increased interest in the history of particular locations. It is at those times that publicity may increase, funding may become available, and numerous projects, including archaeological ones, appear. These tend to generate publicity, publications, and increased community interest. Perhaps such an event has taken place recently in your town, and has sparked your interest in archaeology.

Are these the reasons you decided to study archaeology at this time in your life? Perhaps they are part of the explanation. Other reasons may be more practical. Perhaps a course in archaeology fit a school requirement and your schedule. However, with many options available, you probably weighed the subject matter against that of other courses to help you reach a decision.

Still other reasons may be deeply personal. For example, you may have become interested in archaeology because you were disturbed about the loss of a special site in your home town. Maybe this stimulus has sparked an interest in a career in archaeological—or cultural—resource management or protection! A desire to know what to say, how to say it, and how to argue convincingly for the protection of human heritage in the world may be your justification. Or you may believe that becoming an archaeologist might be a thrilling career choice.

A preliminary question raised by many people, especially students and their parents, has to do with career choices and the best training for them. What do you do with a degree in archaeology? To me, the answer is exciting. Archaeology is part of anthropology, the study of people and their cultures, past and present. We are becoming an increasingly global society, and what could be more relevant in this day of interaction with people from all over the world than to learn all we can about ourselves and our ancestors? When looked at that way, it is clear that this branch of the social and behavioral sciences, as part of a liberal arts education, is a useful direction to take.

To be more specific, a person with a bachelor's degree in anthropology, with an emphasis in archaeology, is a person with a broad educational background. Jobs that specify a bachelor's degree are open to you. If you want a career with more emphasis on the archaeology part of your training and interests, advanced degrees such as a master's degree or a Ph.D. may be useful to you eventually. On the way to those goals, though, there are opportunities for you to gain field and laboratory training and experience—chances to acquire a wide range of knowledge, insight, and "seasoning."

Sometimes field and laboratory experience will count toward college (academic) credit; occasionally there is the chance to find paid employment, even as an undergraduate student, if you have taken an archaeological field school course, laboratory course, or museum-study course, or if you have participated in some sort of internship program that offers these types of experiences. While many of the initial jobs will be relatively low-paying, they tend to open doors to more lucrative opportunities if you are enterprising and hardworking. Often,

Figure 1-1 Ruins of a plantation house near the Smithsonian Environmental Research Center, Ann Arundel County, Maryland. Note the wall thickness, fireplaces, bricks, and chimneys.

Figure 1-2 At Grindstone Island, New York, remnants of cut stone in an abandoned granite quarry are clues to an important late-nineteenth-century industry on the island.

when these jobs are connected to colleges and universities, there are incremental raises and advances in the levels of employment and experience. There may be opportunities to defray costs of graduate work. When the jobs are connected to private contracting companies, there may be similar options, and possibly grants or scholarships to further your training academically—and professionally.

What are some of the job options? After a season of field school training, you may find work as an archaeological crew member, or possibly as a crew chief, helping to supervise other archaeological crew members. Instructional and interpersonal relations skills, as well as knowledge of computers, photography, mapping, drawing, botany, geology, zoology, chemistry, or other areas, may lead to added options within archaeology or peripheral to it. Writing skills, managerial aptitude, and knowledge of or interest in environmental law may lead to other potential career directions.

Thus, when asked what may be done with a career in archaeology, I offer the following preliminary list:

1. Teaching and museum work—All levels are possible, depending on interest and educational degree and certification obtained; age levels for your focus of training may be a factor.
2. Peace Corps or Student Conservation Association—Various levels, places, interests, and skills are available.
3. Excavation—As noted previously, this often is connected with cultural resource management and contract archaeology projects; levels are dependent on training, experience, and area of the country or world; it may include participation in major research projects in locations like Africa's Olduvai Gorge or Hadar Region, Turkey's tell sites, Italy's classical

sites, Latin America's Mayan, Aztec, or Inkan (Incan) areas, or North America's Mesa Verde, Chaco Canyon, Cahokia, or Jamestown sites, or California's Spanish Mission sites.

4. Environmental law—Often this area is connected with contract archaeology and cultural resource management; jobs may be found in local, state, and federal agencies and may require some of the general tests they offer for levels of job placement; legal training may be an offshoot of combined interests in archaeology and laws relating to it.
5. Forensics—The application of training and skills from archaeology, anatomy, physiology, and criminology may lead to this specialization; jobs tend to be connected with local, state, or federal agencies.
6. National Park Service—Various levels and places available that involve different interests and skills; governmental testing may be part of the placement process; park resource management, park curation, supervision of staff as well as site rescue and stabilization after forest fires, floods, or other disasters, are potential job responsibilities.
7. Historical societies—All levels of training and interest in historic and prehistoric human activities, curation training, museology, and other skills may be needed.
8. Tourism—This could involve training for tour-leading to visit various archaeological sites around the world and working with various international organizations; often a range of foreign language skills is essential; additionally, some of the recreation and tourism programs in academic settings may be needed or desirable.

These are some of the potential directions to explore if you are interested in an eventual career in archaeology. Other directions are as varied as your imagination. You may be an exceptional writer of grant proposals, nonfiction articles and books, or fiction with an archaeological focus or theme.

Sometimes there are jobs available with magazines, publishing companies, governmental organizations, local chambers of commerce, and other agencies for a person with combined expertise in writing and archaeology. By now, you may have other ideas that are not listed and you may want to visit your school's guidance or career planning and placement office to seek out other possibilities! Do it!

NOT MY COLOR, NOT MY KIND

Lessons in Race, Class, Age, and Gender in the Academy

By M. Jean Harris

FULL CIRCLE

The scene: Spring 1999, outside the gymnasium-turned-commencement center at a midsize community college in the Pacific Northwest. I am garbed in a black faille gown, a thin, paperback novel secreted within the wide, velvet-striped sleeves of the gown, six-pointed crushed velvet cap firmly on head. Buoyed by a capricious wind, the blue and white hood on my back billows upward as if to join the white-cloud-kissed blue sky. Prodded by patient staff members, the lines of graduates and faculty slowly take shape. Suddenly, oblivious to stentorian commands to "Line up! Get in line," a slim, dark-haired woman breaks formation and runs across the divide between faculty and students to stand looking at me. "Wow," she says. "You look so good, like you were made to wear that gown!"

My former student in Introduction to Cultural Anthropology does not know this is my first opportunity to robe up since receiving the diploma that attests to my having "completed the studies and fulfilled the requirements of the Faculty for the degree of Doctor of Philosophy." She cannot know and there is no time to tell her how pleased I am at her compliment. Instead, I smile and squeeze her hand and respond to her comment about the gown. "Thank you. It does look pretty good, doesn't it?" Enveloped by familiar strains, marching forward, we create pomp and circumstance.

I did not participate in the 1998 commencement ceremonies of the University of North Carolina at Chapel Hill (UNC). By the time of that commencement, I had earned tenure at one college in the Washington State community-college system and moved to teach at another two-year college closer to my home. The nine of us who in 1987 entered the doctoral program

M. Jean Harris, "Not My Color, Not My Kind: Lessons in Race, Class, Age, and Gender in the Academy," *Women in Anthropology*, pp. 163-172. Copyright © 2009 by Left Coast Press. Reprinted with permission.

in the Department of Anthropology at UNC did not graduate as a cohort. I counted the climax of my career as a Carolina anthropology graduate student as the moment in April 1998, when my chair ushered me into the presence of my elder daughter and waiting doctoral committee members to hear the sweet words, "Congratulations, Doctor."

THE BEGINNING

The scene: August 1987, Seattle, Washington. With my fifteen-year-old daughter and two friends, I leave Seattle in a two-car caravan consisting of my aging Chevette and a "drive-away" car to be delivered to a student in Chapel Hill, North Carolina. I have traded a supervisory position in the human resources department of a local public utility for that of graduate student in anthropology. Our destination is an apartment I have leased, sight unseen, in Durham, North Carolina.

We arrive in Durham on an August day so hot the blacktop road is sweating. Bone-weary from the long drive, I glimpse our new home from the freeway but cannot find the way to it. I enter and exit the freeway again and again. Finally, near tears, I take a deep breath and exit yet again, and drive down one sun-bruised, deserted street and then another, relaxing now into the twists and turns in the road. One more turn, then there before us is the sign at the entrance to the pseudo-columned Colonial Arms Apartments. Looking back, my route from the freeway to the apartment is emblematic of my time in graduate school: much time was spent trusting intuition to find my way to a goal I saw but had no map to reach.

MY EARLIER ACADEMIC CAREER

Prior to enrolling in the graduate program in anthropology at UNC, I had always been a successful student. Mama, my grandfather's sister who was the mother who raised me, had only gone through fifth grade, but she was a willing listener to the events of my school day and she corrected my homework assignments until they exceeded her skill level. From that point, Mama monitored my activities to ensure that I completed the work. Skilled in common sense, Mama early on explained to me that I, as a black person, would not always receive the grade I had earned, so I should work to satisfy myself. Consequently, though I received good grades, I did not put much stock in them. Grades were subjective, awarded by mercurial teachers, but standardized tests had meaning. Teachers themselves responded to standardized test scores, responses scored in distant locations without regard for responder. I came to believe that standardized tests were a fair measure of what I had learned. In elementary school, I tested into the gifted program; in high school, I was placed in the college-prep track. During my senior year of high school, I worked twenty hours a week after school; earned As and Bs; and, based on standardized test scores, earned a national scholarship to assist with college tuition.

The autumn after graduation, in the fall of 1961, I enrolled in a small, academically respectable, sectarian college. Supremely confident in my abilities and preparation, I turned a deaf ear to doomsaying faculty who in orientation told us, "Look to your left, now look to your right. One of you won't be here next quarter." I knew *I* wouldn't be the one who was gone! I believed that I could compete with students from the "better" (read "more white") high schools. Although I encountered instructors who initially discounted my abilities because of my race, I was right. My confidence was confirmed. By the end of my freshman year, I was eagerly sought out as a participant in campus colloquia. I still worked twenty hours a week and full time during breaks and summers.

Along the way, I discovered that courses at the nearby state flagship research institution were easier than at my college because the classes were so large that instructors administered multiple-choice tests. Periodically, I attended this university for a quarter's worth of "easy" credits. Away from home, away from my fundamentalist evangelical church, I joyfully balanced academic and social life. At the end of my junior year, I was inducted into the women's scholastic and service honorary. I worked a full forty-hour week during my senior year, yet my performance in independent reading courses encouraged two professors to advise me to consider graduate work.

I did not enroll in a graduate program immediately after college. Instead, I gave birth to a baby girl and went to work full time. After working for two years, I was accepted into the social welfare program at the University of California, Berkeley, and moved to California. For the first time in my life, I was studying in the company of a number of black students. Our first-year class was nearly 25 percent black. I was stretched thin by my workload—twenty hours of work a week, a full load of classes, and parenthood—but I was prepared for coursework at what I had heard was the most intellectually demanding university on the West Coast.

The heady, golden season of challenging professors and challenging study groups ended when my fiance returned from his tour of duty in Vietnam, not to San Francisco as we had hoped, but to Tacoma, Washington, near my home and my seriously ill Mama. I transferred to the University of Washington, where I completed my Master of Social Work degree without difficulty. Marriage, the birth of another child, the death of Mama, and then a divorce delayed my half-formed plans to use my social work degree to finance graduate study in sociology. Still, I believed that I would someday become a graduate student in a doctoral program. I did not return to university for more than fifteen years.

SHIFTING GEARS

Like most of the black women I knew, I had combined marriage, motherhood, and full-time work in a variety of settings. At the time of my divorce, I had obtained tenure at a local community college and gone on to a position as program evaluator at the State Board for Community Colleges, a job that required extensive travel. Faced with the logistics of arranging childcare with a heavy travel schedule, I obtained a job in the private sector and taught part time. I continued to explore possibilities for graduate study.

My older daughter exited college as the younger entered high school. The time seemed right for me to return to university. Friends introduced me to an anthropologist couple who encouraged me to explore anthropology, "the least restrictive of the social sciences." Carefully, I plotted my future. I used my degree in social work as leverage to enroll in a graduate course in tests and measurements, and used myself as subject for the course. The tests indicated that my interests matched those of college professors.

The process I used to select a graduate program was unorthodox. I did not evaluate possible programs based upon whether there were professors in my fields of interest. Instead, because I would be moving with a fifteen-year-old, relatively naive daughter, I eliminated programs in major metropolitan centers. Because I was a Northwesterner, I eliminated programs in locations that had extremes of climate. I then checked the ratings on programs that met my criteria and applied to those on my short list. I took the GREs, earning a score high enough to receive an honorable mention for a National Science Foundation fellowship and access to the foundation's computing center. My final choice of program was based upon a combination of my impression of the school during a campus visit and the financial-aid package. I felt confident I would complete the program in anthropology in time to be able to finance my daughter's tuition. With high hopes, I embarked on what proved to be, up to that point, my life's most painful journey.

CULTURE SHOCK

Shortly after I arrived in Durham, the university commenced orientation for autumn term. Faculty in the Department of Anthropology told our class, comprised of four white men, two white women, two international students, and me, that instead of admitting a large class with the expectation that a substantial number would wash out, they had selected us nine with the expectation that each of us would finish the program. This orientation certainly sounded better than the one I'd had as an undergrad! I recall thinking that graduate school was part hoop jumping, part endurance exercise, and part intellectual endeavor.

Initially comforted, I soon found my confidence eroding. My fellow students were, for the most part, the offspring or relatives of academics or other high- status professionals. Suddenly self-conscious, I felt I did not fit. I had returned to school after a long absence. My North American classmates were young recent graduates of selective colleges and universities. My professional training in social work and my work experience reinforced the use of the plodding, ponderous language favored in business and government bureaucracies. My fellow students had honed the elegant turns of phrase rewarded in academia. Several years had passed since I had last taught. There had been a paradigm shift—no more functionalism; Marx reigned! There I was—thousands of miles from home, far from friends who believed in me, desperately trying to support myself and my daughter on a stipend meant for one person—paralyzed by self-doubt.

I felt intimidated, terrified that I was inadequate to the rigors of graduate school, and certain I had ruined my miserably alienated (later diagnosed as clinically depressed) daughter's life. Unable to borrow enough money to live on, unable to hire a lawyer to obtain more child support for my daughter, I took a job as a part-time census enumerator, then felt torn because I had no time for my daughter, no time for my studies, no time for myself. It did not once occur to me that the graduate school program was not structured for a person of my age, class, race, gender, and marital status.

Although my performance the first year was more than satisfactory, I was thoroughly miserable and alienated from most of my fellow students and our professors. I learned that some faculty had reservations about admitting me and another mature student (their age-mates!) because older students were "not flexible." My daughter changed from an outgoing, enthusiastic, self-reliant student to a silent, sluggish waif. Without the self-monitoring skills that had buoyed my confidence in past times, I had no idea why on one paper I received an excellent grade, and on another merely an "adequate, light on analysis." Suddenly, I, who had been named copy editor of the high school creative-writing magazine, editor of an edition of the college creative-writing magazine, and yearlong editor of the student newspaper; I, who counted on grades in English to elevate my GPA, no longer understood how to structure an essay. I, elected in college to a national speech honorary, became tongue-tied. In a yawning divide where common understandings fail, one of my favorite professors told me that she was perplexed by my lack of interest in my work, and I, mortified, found no words to tell her that I was thoroughly interested but confused and utterly overwhelmed.

Class, age, gender, and race intertwined to form a multistranded rope of misery. A professor for whom I was a teaching assistant insisted that I come to her house to grade papers. I resisted because doing so brought back memories of accompanying my Mama to do "day work." A senior professor in the department introduced me to another graduate student, a young white woman. He introduced her to me as "Ms. So-and-so" and identified me to her by my first name. I was very aware that he was among a minority of professors, "Southern gentlemen," who practiced similar asymmetry of address when conversing with or identifying the black secretary and the white senior secretary. This Southern gentleman looked put out when I asked the student whether she wanted each of us to address the other formally, by title and last name, or informally, by first name.

I was likewise incensed when the same professor condescendingly told a female graduate student that it was about time she made some more cookies for the office. The giggly response made by this highly competent woman and her prompt compliance with his request helped me understand why I was viewed as not having been "raised right" as either a black person or a woman. At that moment, I was too involved to realize that I was encountering situations exacerbated by differences of regional culture, differently gendered upbringing, and different understandings of education as a merit system.

Well-intentioned but culturally insensitive advice further fueled my sense of alienation. Faculty members, sincerely trying to be helpful, touted the desirability of the public high school in Chapel Hill as the best school for my daughter. Only one professor understood that considerations other than SAT scores and extracurricular offerings could affect my choice of high school for my daughter. He provided me with the telephone numbers of black parents, who informed me of long-standing and intense strains between black families and the local high school. Likewise, an opportunity to live in the rural part of the county was a selling point to attract white students to the program, but black students told me that a single black woman not from the region probably wouldn't want to live in the county because of recent Klan activity.

Out of a desperate desire to improve my financial situation, much as one might buy a lottery ticket, I applied for a Ford Foundation graduate fellowship for individuals who were members of minority groups underrepresented in higher education. At about the same time, I enrolled in a community creative- writing class at a branch of the Durham Public Library. Faculty members were skeptical regarding my chances for the former and gravely doubtful regarding the usefulness and advisability of the latter. My advisor and the only other black American student in the department encouraged me, though they admitted that they sometimes did not understand my misery. During a long, solitary walk, I recalled an article written by Douglas Davidson, a sociology graduate student during my days at Berkeley, "The Furious Passage of the Black Graduate Student."[1] I located the article; Davidson voiced many of my frustrations. Slowly, I became less anxious and more able to think and write. Near the end of the first year of the program, I learned that I had won a three-year Ford Foundation fellowship! My advisor visibly relaxed as I informed her of the award. Her hug and forceful expression of joy at my good fortune made me know that she had felt both my pain and the frustration of the department at my maladjustment.

REFLECTIONS

Mine is not a story with clear villains or heroes. It does not have a fairy-tale ending. My reflections upon my time at Chapel Hill contain nearly equal parts joy and pain. In the midst of misery, I formed friendships among faculty and students that will be mine for the rest of my life. I reconnected with friends from my days at Berkeley who provided me with tangible support in the form of references, offers to read and comment on my research proposal, weekends away from North Carolina, and the equally necessary intangible emotional support. Recently, my elder daughter died from cancer. Members of the department called, wrote, sent flowers. Friends from Chapel Hill came to see us. One, stretching the limits of telecommuting, stayed with me for three weeks. Nonetheless, when I recall my graduate experience, my eyes still fill with tears. The assault on my sense of self, the intellectual insecurity are deep, slow-healing wounds that may color the rest of my career.

The value of reflecting upon my experience lies, in part, in analyzing my case to see how my experiences, though uniquely mine, are similar to and different from those of others like me. I am, to paraphrase an anthropological saying, like all other mature women who returned to school to earn a doctorate, like some other mature women who returned to school to earn a doctorate, like no other mature woman who returned to school to earn a doctorate.

The intertwining factors of age, gender, and race shape my view of the nearly eleven-year journey from graduate student to Doctor of Philosophy, but my perceptions are especially influenced by issues of class. I know well from black feminist theorists the folly of attempting to separate or privilege multiple oppressions like age, race, gender, class. If I keep in mind that, in my experience, these factors function always together, I can discuss them sequentially.

Observers from older cultures—from Frenchman Alexis de Tocqueville, whose *Democracy in America* was first published in the 1830s, to Chinese American anthropologist Francis Hsu's research in mid-to-late-twentieth-century America—have commented on the youthful brashness, exuberance, and informality of Americans. Our nation and culture are relatively young, and we value youth. In our fast-moving, innovative society, the knowledge of older people is not always immediately applicable. It is perhaps an easy step to equate being older with being obsolete. Certainly, that is true with technology; perhaps it is true with people, too. Although we are living longer as a society, our general expectation for developmental tasks for members of my generation was that young adulthood was the period when one "completed" one's formal education and settled upon a career. Returning students, those reentering formal educational programs in their late twenties or after, were somehow out of step with their fellow students. Perceptions of less time to contribute to a career could jeopardize admission to professional programs such as medicine. Within the academy, perhaps there was greater recognition that mature, even "old" scholars could make significant contributions, though generally such mature scholars had still been relatively young when they made the initial contributions that earned them the doctorate and admission into the academy. Late-entering graduate students were suspect. They had no track record. How could one evaluate their "promise"? Today, in the universities in my state, and I suspect in many universities elsewhere, we give lip service to life experience, but not much college credit for it. During my graduate experience, examples from life were minor currency, elaborations on scholarly concepts. Thus, I am sure, being older was no asset.

Apart from learning that my age and that of my fellow classmates were mentioned when we were being considered for admission, I have no evidence that my age may have negatively affected my status in the program, though it was true that my having a daughter who was close to most of my classmates in age made for a bit of awkwardness in social interactions. Some students partied with both of us, some with one or the other. When we were out drinking, sometimes faculty seemed unclear whether I was in a peer or subordinate relationship. Moreover, I had more-varied work experience than most of my professors and both less fear of and less tolerance for the traditional, ritual debasement of graduate students. Perhaps, too, because I was older, I was somewhat less awed by my professors and more likely to think them merely eccentric, rather than taking their eccentricities as evidence of brilliance.

Gender intertwined with age. The most senior faculty men seemed to establish rapport and mentoring relationships with men students more easily than with the women. Traditional role expectations for women were present in everyday interactions, from the earlier-mentioned incident regarding cookie baking to opportunities to earn extra money by providing childcare for faculty offspring. The unfortunate experience when I was expected to grade papers at the home of a professor would not have happened had I not been a woman. No male student ever had that experience, although other female students did. I am likewise certain I would not have had the experience had I been an African American man of my age.

Race. Race—the always-present, seldom openly-talked-about factor in the experience of people of color—permeated my experience at Carolina. When I entered the graduate program, the UNC system was still under a consent decree because of race issues. The anthropology department in its long history had never graduated an African American man or woman with the PhD. My classmate from South Africa was the first person of African descent to finish the program; I am the first African American.

The all-campus black graduate student organization included students in all disciplines, yet we were few in number. As one of my part-time jobs, I was hired in a program to provide support to African American undergraduates. I heard them lament many times that they were told, directly or indirectly, by students and faculty that they were only admitted because of affirmative action. The implication was clear that they were not perceived as intellectually qualified to be at the flagship research institution, a perception that I did not share based upon my experience as a teaching assistant in two different departments. Because African American undergraduates were the same race as I, I imagined that they could be my children enduring painful experiences, and many times I was thankful that I was well matured before I encountered the condescending, patronizing attitudes that were rampant on campus.

Before I went to North Carolina, I had been a successful personnel professional. Trained as a recruiter at a Fortune 500 firm, I had managed the recruiting function in a large, multiple-branch financial institution and had coordinated the employee assistance program at the same corporation. Immediately before enrolling in graduate school, I had supervised the employment department for a regional public utility company. I had excellent references, so I attributed it to racial factors that I was never called for even an interview for the numerous personnel jobs at various levels that were available at the university and in the town (I did not remove from my resume such racial identifiers as professional memberships and volunteer experience).

Although I did not know all of the African American students, faculty, and staff on campus, there was an imagined community that allowed us to approach each other more easily than we did students of other "races." When I first visited the campus, the department introduced me to the only other African American graduate student in anthropology. Our relationship of mutual support was cemented by our racial experiences in the South; in our case, race transcended class and regional culture differences.

Multiple factors influenced my graduate experience, yet I recall class factors most vividly. Bowles and Gintis[2] and MacLeod[3] provide analyses of the role education plays in reproducing a class structure. In his writings, French anthropologist Pierre Bourdieu deals extensively with cultural competence, cultural pedigree, and cultural capital. Bourdieu, along with Marx and Weber, provided tools for me to apply to my situation. Had I had the objectivity to consider my social class before I enrolled in graduate school, I might have been far less traumatized, for I would have understood the limitations of my education to that point. Had I not suffered from "false consciousness," I would have understood that the notion that education is a meritocracy obfuscates its support for the status quo. I would not have been surprised that I, more than the other grad students, tried to give professors what they expected of me. I would not have been surprised that the rules were different in the arena I had entered. I might have understood that the knowledge that is counted on in standardized tests is not sufficient to succeed at the level I had entered, and I might have more confidently engaged the assignments. Surely, I would have understood which productions were rewarded, and why.

Workers are not rewarded for original thought. Working-class people enter the academy in spite of their class position, not because of it, and in general, they shed working-class thinking, working-class behavior, working-class loyalties if they are to succeed. In Weberian terms, I was not working class. The class I was raised in and identified with was clearly a lower class. I was raised "on the dole" and had long outgrown feelings of shame about that fact. Moreover, I strongly identified with the substantial numbers of working-class students I taught in community college and wanted them to gain tools to earn better wages. I was irritated when my professors gently countered my plan to earn a doctorate so I could return to community college teaching, with suggestions that my ideas would change as I realized I could "do more." No wonder, too, that the professor with whom I most strongly identified and who most strongly supported me was a white woman whose class of origin was more similar to mine.

My preferred communication style, relatively direct and confrontational, and my emphasis on pragmatic applications of theory all marked me as Other, but worst of all, my bullheaded resistance to changing my style made me suspect. Either I did not know the rules or I was deliberately refusing to play. In any case, perhaps I should not have been allowed to join the host in the ivory tower.

My research interest, African people in diaspora, especially African people in the United States, was not yet fashionable in anthropology when I entered graduate school. Neither were my synthesizing, interpretive approach and my interest in being accountable to my ethnic community. I recall my chair's dismay when I told a member of my committee that my greatest fear was that black people would say my work was "some esoteric, peripheral bullshit." I did not know where I was off, and most of my professors could not articulate what was wrong. Mutually frustrated, I became increasingly alienated—and they began to doubt my potential. Had we recognized how my class background was influencing my experience, I would have made choices with more awareness of the implications and consequences. Who knows, I might have taken a different path entirely.

Cultures clashed during my graduate experience, and I was changed. I am neither the person who entered graduate school nor the person who exited. I cannot weigh the factors that influenced my graduate experience separately, and I cannot at this time guess how they interacted with each other. I do know that I should be less surprised that I had challenges than that I surmounted them. Education is nothing if not change, a "leading out," in the sense of the Latin from which our word *education* comes. Anthropology provides us with tools to understand ourselves and the Other. Despite the pain, I have met both goals.

ENDNOTES

1. Davidson (1993).
2. Bowles and Gintis (1986).
3. MacLeod (1995).

REFERENCES

Bowles, Samuel, and Herbert Gintis. 1986. *Democracy and Capitalism: Property, Community, and the Contradictions of Modern Social Thought.* New York: Basic Books.

Davidson, Douglas. 1993. The Furious Passage of the Black Graduate Student. In *The Death of White Sociology*, ed. Joyce A. Ladner, 25-51. New York: Random House.

MacLeod, Jay. 1995. *Ain't No Makin' It: Aspirations and Attainment in a Low-Income Neighborhood.* Boulder, CO: Westview Press.

GARBAGE AND GARBAGE-IN-WAITING

By Barbara J. Little

It takes very special qualities to devote one's life to problems with no attainable solutions and to poking around in dead people's garbage: Words like "masochistic," "nosy," and "completely batty" spring to mind.

—Paul Bahn

What would archaeologist Paul Bahn say about poking around in living people's garbage? Archaeologists have not been content for our skills to be confined to the study of the past. One part of modern material culture studies is the study of modern refuse dubbed "garbology." William Rathje and his students started the Garbage Project in the 1970s to use archaeology to address contemporary concerns, such as questions important to modern waste management and sociology. Garbologists have succeeded in measuring diet and nutrition, assessing resource waste and proposing methods to minimize waste, evaluating household participation in recycling programs, identifying household-level sources of hazardous wastes, cross-validating census counts of minority populations, and providing base data for the design of new "environmentally friendly" packages.

The Garbage Project rests on the simple premise that if discards can provide important information about past societies, then they should also provide such information about societies today. For the first couple of decades of this ongoing project, researchers sampled and recorded household trash (always guarding the anonymity of the residents). More recently they've expanded their work to landfills and exploded common myths about our solid waste problem.

Studying fresh garbage from households is messy and smelly. The garbologist-archaeologists sort and record packaging, food, and other waste items into myriad categories, including separating "once-edible food" from "food preparation debris." They have compared their data with surveys that record household responses to questions about what they throw away.

Analysis reveals some basic patterns in people's actions and perceptions. First, what people do is not the same as what they say they do. People do not accurately report what food they waste, what they recycle, their hazardous wastes, or what they eat and drink. Interestingly enough, there are some regularities—or, as Rathje terms them, "syndromes"—in the ways that people misrepresent their own behavior. The "surrogate syndrome" refers to accuracy provided by someone other than the consumer. For example, drinkers underreport their alcohol consumption. However, nondrinkers in a household accurately report what the drinkers consume. The "good provider syndrome" refers to the tendency for adult women reporting for a household to over report everything the household uses.

Second, people who respond to surveys report rational behavior, but their actual behavior appears to be irrational. For example, during the United States' nationwide beef shortage of 1973, consumers complained about the high cost of beef and yet discarded more edible beef waste than ever. The additional waste probably was due to people buying cheaper and unfamiliar cuts, perhaps not knowing how to best prepare them and disliking the result. More recently, in response to admonitions about healthy diets and the need to consume less fat from red meat, people did indeed cut and discard more fat from their beef. However, they also consumed more processed meat, like sausages and hot dogs, which of course contains more fat and is less healthy.

From 1987 to 1995 Garbage Project archaeologists studied 30 tons of deposits from 15 different landfills across North America. Conventional wisdom suggested what they would find, based on popular assumptions about waste and biodegradation. Predictions were that fast-food packaging and disposable diapers would predominate; instead these items made up only 2 percent of the volume. Surprisingly, items that had received little public attention turned out to take up a lot of space. Even though researchers tried to avoid drilling through debris from construction and demolition (because of the damage caused to their equipment), such debris made up 20 percent or more of the landfill volume. The largest category, however, was paper. Nearly half of all the landfill that researchers excavated, measured, sorted, and recorded consisted of newspapers, magazines, packaging, printouts, and phonebooks.

Since the studies, many communities with recycling programs have understood the need to keep paper out of landfills, as it doesn't biodegrade as well as hoped. One of the resulting recommendations for all of us is that we should buy goods and packaging with "post-consumer recycled" content.

One of the general findings of the Garbage Project, when the results of many years of research are considered, is a somewhat disturbing disconnection between what people think is happening and what is really happening. Rathje offers an important observation from his decades of research about apparent disinterest in the material realities of behavior. In spite of some behavioral changes in response to findings (like changing recycling programs for paper), he reports that many of the Garbage Project's studies have been ignored after the results are reported because the results do not match preconceptions or other interests. He has surmised from his experience, for example, that the alcohol industry really doesn't want to reveal the extent of alcoholism in the United States. He also is surprised at the apparent disinterest from the U.S. Department of Agriculture in real data on household nutrition.

Rathje (2001:68) remarks: "I used to believe that people intentionally 'lie' in interviews today—in the past in texts and monuments. Now, I wonder if the differences between mental and material realities are not mainly simple cases of people not mentally recognizing material realities and fooling themselves." Along with Rathje, many historical archaeologists have direct and often frustrating experience with people's separate realities of mental, behavioral, and material life. Many of us observe disjunctures between the relative meaning and authority granted to different kinds of evidence, as documents are often assumed to be true even when physical material evidence provided by archaeology contradicts it (not what we may hope for, given our legal system's emphasis on material evidence). The relationship among such different realities is, as Rathje explains, a very important question for behavioral science to address because "only when all of these realities are in sync with each other can people plan rational public policies and make rational personal decisions." In other words, we may not know what we think we know and those misperceptions hurt our ability to make rational choices with predictable results.

Faced with obvious and important gaps in our understanding, of course, we can try to learn more and gain some useful understanding. One of the ways we learn is to design research by coming up with questions and then figuring out what kinds of evidence are needed to begin to address them. Consumerism clearly is directly related to garbage; we might even think about the things we buy as pregarbage. In the remainder of this chapter, I simply want to direct you to some issues and questions to think about. They come from the work of Teresita Majewski and Michael Schiffer, who study consumerism to better understand our modern society. They identify the following series of issues and propose questions that highlight the present-day relevance of material culture studies.

Issue #1. Structural and behavioral aspects of the emergence, growth, and maintenance of consumer societies. What do we need to understand to address this issue? We need to understand economic structural factors; the legal system and its effects on producing, selling, and buying; and both written and unwritten social rules about appropriate ownership. We need to ask how an activity becomes consumerized. That is, how can an activity come to require a continual updating of the material culture that's perceived as necessary for the activity to continue? Consumer societies are resource and energy intensive, but how is the concept of sustainability coming into play? How have international aspects of manufacturing, marketing, and consumption reconfigured people's lives? What are the roles of huge corporations in these processes?

Issue #2. Effects of consumerism on the life histories of specific products. Think about consumerism's driving force of novelty and frequent replacement of goods. For example, when and how does novelty become equated with and valued as progress? How does an annual "need" for new models of products become established and successfully marketed? How does the need for novelty get spread to more different kinds of consumer, commercial, and industrial goods?

Issue #3. Advertising and communication. Questions here involve mass-communication media. How are these media used to maintain and spread such values as the "novelty orientation," which appear to be necessary for present-day consumerism to function?

Issue #4. Explaining apparent alternatives/reactions to consumerism. Even within a consumer society, there are alternatives to rampant consumerism. How do people develop and practice resistance strategies against consumerism? How do people use artifacts to create meanings and identities? How are alternative lifestyles maintained? It is also worth asking how such resistance might be co-opted by the larger society. Specifically, Majewski and Schiffer

(2001:33) ask, "How and why do handcrafted products of traditional societies become integrated into the consumption processes of industrial consumer societies?"

Issue #5. Commercialization process of consumer services and societal practices. Among many questions here are those about how religious items and practices and personal services (such as grooming and medical care) become consumerized.

Issue #6. Ideological expressions of consumerist societies. Ideological expressions provide a particularly rich, if somewhat ambiguous, source of questions for material culture studies. I like the way that Majewski and Schiffer state these questions:

> When and how did the ideology of science and technology as founts of wondrous new products become entrenched? How is this ideology related to the artifacts purchased by middle- and working-class people over the last century? How is the erection of monumental architecture in cities since 1850 related to changes in the relative wealth and social power of churches, polities, various kinds of corporations, and sports franchises? (2001:33)

Archaeologists believe that archaeology contributes an important methodology that can address such questions. In thinking about the material culture that we own, or want to own, it is worth some pondering about these issues and questions as they relate to our lives and choices.

DEATH OF AN ICON

By Jeff Alexander

> When you see loons dying in large numbers it's very sad.
> You're losing that voice and symbol of the wilderness.
> —Ward Stone, New York state wildlife pathologist

The sun was sinking toward the horizon over Lake Ontario, its fading light painting the autumn sky orange and lavender, when the icon of the North filled the air with its haunting yodel. The bird broadcasting the call was not visible from the beach at Henderson Harbor in northern New York. Still, there was no mistaking that a common loon, *Gavia immer,* was nearby. The loon had one of the most distinctive calls in the community of water birds. Its memorable call was matched only by its beauty and affection for its offspring. With monochrome feathers, crimson eyes, a black-and-white-barred necklace and checkered pattern on its back, the common loon struck a distinctive pose as it cruised low in the water. The gorgeous, symbolic birds were known to carry their chicks on their backs to protect them from predators. They were anything but common. Among water birds of the North, the common loon had no equal, in my estimation.

So revered was the common loon, its likeness adorned the Canadian one- dollar coin, the so-called "loonie." It was Minnesota's state bird and a symbol of the wilderness that still reigned supreme in northern boreal forests. Its plaintive wail and laughing tremolo were the sounds by which many who visited the northern Great Lakes and Canada measured their wilderness experience. "Loons ... provide much of the magic of the northern wilderness. Their haunting calls become the raw material for our own dreams and imaginations," said writer Judith W. McIntyre, who spent 20 years studying the birds. Hearing a loon was a reliable indication you

The common loon, with its distinctive markings and haunting calls, is an icon of wilderness in the Great Lakes region. Source: U.S. Fish and Wildlife Service.

Invasive mussels and fish, imported to the Great Lakes by transoceanic freighters, passed a strain of botulism up the food chain, killing tens of thousands of loons and other water birds. This bird was one of many that washed up at Presque Isle State Park in Erie, Pennsylvania.

had escaped the cacophony of city life and found a measure of solitude. It was the beauty and symbolism of these birds that made it so disturbing to see them washed up, dead, on Great Lakes beaches.

In November 2007, I went looking for birds I did not want to see. Dead loons were washing up on the eastern shores of Lake Ontario. It wasn't an isolated problem. More than 8,000 dead loons washed up on Great Lakes shorelines between 1999 and 2007, their carcasses casting a pall over hundreds of miles of otherwise scenic beaches along Lakes Erie, Ontario, Huron, and Michigan. Sadly, the shoreline of Lake Ontario's Henderson Bay, about 15 miles southwest of Watertown, New York, was but one of a growing list of beaches that became death traps for loons in the first decade of the twenty-first century.

Irene Mazzochi, a biologist for the New York Department of Environmental Conservation, allowed me to tag along on her weekly check for dead birds on the coast of Henderson Bay. The picturesque beach fronted the state's Black Pond Wildlife Management Area and El Dorado Shores, a Nature Conservancy preserve that featured some of the most spectacular sand dunes in the eastern Great Lakes. It certainly wasn't the kind of place one would expect to find bird carcasses littering the shoreline. Mazzochi was driving a state-issued pickup truck along the beach, a death patrol carried out 30 feet from the water's edge, when one of her assistants sitting in the back pounded on the rear window.

"Bird," he yelled. Mazzochi hit the brakes and turned off the truck's engine. Everyone piled out and walked somberly toward the water. "Good eye," she told her colleague. He had spotted a dead loon rolling on its chest in the tiny waves that lapped at the shoreline.

Everyone knew the drill: Two workers in rubber boots and latex gloves pulled out a large, clear plastic bag; a third put on a pair of latex gloves and picked up the dead bird by its webbed feet and dropped it in the bag. The bag was placed in the back of the truck for transport back to a field station. From there, it would be sent to a laboratory to confirm, or dismiss, the presence of avian botulism. Then the carcass would be incinerated. The bird was the third of four dead loons and five deceased long-tailed ducks we found on the beach that raw November day. Our haul was a pittance compared to what Mazzochi's crew had found the previous week. In the second week of November 2007, the documenters of death turned up 66 dead loons and 41 other dead birds on that same stretch of beach. "When we drove out on the beach, everywhere you looked you could see white spots—those were the loons," Mazzochi said. The dead loons often were found lying on their backs, wings clutched tight to their bodies, their snowy chests facing the heavens.

A spirited Italian woman working in the male-dominated field of wildlife management, Mazzochi said she considered removing dead birds from the beach one of the worst parts of her job. Still, it was important work that had to be done. Leaving dead birds on the beach increased the risk of raccoons and other scavengers ingesting the deadly botulism bacteria. On an emotional level, the sight of dead birds on a beach was depressing. Physically, it was nauseating.

"People don't like to go out to the beach and see dead birds," Mazzochi said. "And if we can keep other animals from ingesting botulism it's a good thing. But this is pretty sad for me, picking up all these dead birds. It's not a glamorous part of my job. I'd rather be monitoring birds in a wetland or doing environmental education for students. I usually spend a lot of time in a canoe in wetlands. I feel like I almost have to get a cold heart when I go to pick up these dead birds. You can't look at every dead bird you pick up and say, 'Oh,'" she said, her words fading to an awkward silence.

Mazzochi's cell phone rang as we talked about the deaths of glorious loons. A colleague called to report several dead loons along the rocky shoreline at Cape Vincent, the town situated where Lake Ontario flows into the St. Lawrence River. "You might want to go up there if you need more pictures," Mazzochi told me. It wasn't too long a drive, and Cape Vincent afforded a chance to see where nearly all water from the Great Lakes funneled into the St. Lawrence River.

I drove north up Highway 12E, a rolling two-lane road that passed through bucolic farmland and the curiously named villages of Limerick and Lyme. Arriving in Cape Vincent, I headed west along the St. Lawrence River. Within minutes, I spotted dead loons. The birds' brilliant white chests stood out like soldiers waving the white flag of surrender. I pulled off the side of the winding road and walked down the rocky bank to get a closer look.

While snapping photographs, I heard a faint, rhythmic sound approaching from the east: Thump, thump, thump, thump, thump, thump. The deep, pounding sound grew a bit louder with each cycle. A minute or so passed before an ocean freighter headed for Lake Ontario emerged from behind a small peninsula. The timing was uncanny. It was freighters like that one, ocean ships from Europe and Asia, that infected the Great Lakes with the foreign mussels and fish believed to be responsible for a chain of events that ultimately killed the loon lying at my feet.

The dead loons on the shores of Lake Ontario and the St. Lawrence River were eerily reminiscent of the scenes Rachel Carson described in her landmark 1962 book, *Silent Spring*. Carson wrote about how the careless, rampant use of the pesticide DDT in the 1950s killed countless robins and other songbirds. The chemical also thinned the eggs of bald eagles, greatly reducing the number of offspring and pushing the spectacular birds to the brink of extinction in the Great Lakes region. Carson's book, which led to a nationwide ban on DDT and launched the modern environmental movement, questioned the wisdom, the arrogance, of humans who decided which parts of the natural world could be sacrificed for the sake of economic gain. She wrote: "Who has made the decision that sets in motion these chains of poisonings, this ever widening wave of death that spreads out, like ripples when a pebble is dropped into a still pond? ... The decision is that of the authoritarian temporarily entrusted with power; he has made it during a moment of inattention by millions to whom beauty and the ordered world of nature still have a meaning that is deep and imperative."

Fifty years after the St. Lawrence Seaway opened, I wondered if a scenario similar to the DDT tragedy was playing out in the Great Lakes. Was the decision to open the lakes to ocean freighters for the sake of commerce carried out by a powerful few with no regard for how it might affect the nature of these freshwater seas? Were the deaths of 70,000 water birds that washed up on beaches over a 10-year period an ecological crisis, or simply part of the cost of foreign freighters doing business in the Great Lakes? Looking at a dead loon lying face up on the shore of Lake Ontario, its haunting call silenced, I wondered if this iconic species was the canary of the lakes. There was a time when coal miners took caged canaries with them deep into mines to gauge air quality. If the birds stopped singing, it was a warning that methane levels were dangerously high and might explode.

Common loons that spent the majority of their lives on remote lakes in northern Canada were an indicator of water quality, fish populations, and serenity. Lakes too polluted to support healthy fish populations, or too heavily trafficked by powerboats or personal watercraft, rarely supported loons. As top predators, loons were a barometer of ecosystem health. Problems in the food chain, caused by toxins or a lack of nutrients, were often reflected in the health of fish, birds, and mammals at the top of the food chain.

When thousands of loons fell silent on Great Lakes beaches, it said something about an ecosystem that provided the birds sustenance during their fall migration. Were these dead icons of the North a warning, much like the dead canaries that foretold dangerous conditions in coal mines? If so, what were the silenced loons telling us about the health of lakes that the St. Lawrence Seaway linked to the global shipping network and all the world's oceans 50 years ago? Let's find out.

HOPE AMID THE RUINS

By Jeff Alexander

> The fact of the matter is we're not driving this bus anymore.
> Hang on for the ride—we're going to see where this process goes.
> —James Baker, Michigan Department of Natural Resources fisheries manager, 2007

Rarely do professional sporting events provide lessons in ecology. Sports are about competition, teamwork, endurance, and athletes overcoming obstacles to achieve greatness. Once in a while, though, athletic competitions provide a stage for lessons that have little to do with strength and determination, wins or losses. Such was the case on an unusually warm October night in Cleveland, Ohio, in 2007. It was October 5. The Cleveland Indians were battling the mighty New York Yankees in an American League playoff game. The stakes were high—the winner of the series would advance to play in the World Series, the apex of America's national pastime. With the Yankees leading 1-0 in the seventh inning, the Bronx Bombers sent in rookie pitching sensation Joba Chamberlain to preserve the lead. Chamberlain was firing 100-mph fastballs at batters when a cloud of mayflies descended upon Jacobs Field. The insects swarmed around the young pitcher's head and clung to his neck, arms, and jersey. Yankees catcher Jorge Posada tried to repel the mayflies by spraying insect repellent on Chamberlain. That backfired—more of the bugs clung to the pitcher, rattling his nerves. Chamberlain lost his focus and control of his pitches. He walked two batters, threw two wild pitches, and hit one batter, which allowed Cleveland to tie the game. The Indians went on to win, 2-1, in eleven innings. Cleveland's baseball fans celebrated. So did scientists and anglers who understood the ecological significance of the swarming mayflies that provided a bizarre home-field advantage for Cleveland's beloved Indians.

The return of mayflies to Lake Erie was something of an environmental miracle. Gross pollution in the 1940s and '50s created dead zones in the lake, areas lacking enough oxygen to support aquatic life. Among the victims were mayflies, which were nearly eliminated from the lake scarce over the next three decades. Then, in the early 1990s, mayflies made a surprising, mysterious, and dramatic return to Erie's western basin. Like a bolt from a blue sky, swarms of mayflies took flight over the lake, delighting ecologists and anglers. The annual mayfly hatch was so thick in 1996, some communities along western Lake Erie posted signs warning motorists of slippery conditions created by the oily remains of mayfly road kill. The massive hatch caused a brownout over northwestern Ohio when the insects coated a power plant's substation. Attracted by the substation lights, the mayflies formed a blanket that was so dense, their tiny bodies conducted electricity across the facility's insulators. The power surge knocked the substation out of service, causing lights to dim across the region.

Most researchers believed that water-quality improvements achieved after passage of the federal Clean Water Act, which increased oxygen concentrations in the lake, spurred the resurrection of mayflies. But there was more to the story. Mayflies returned so suddenly, and in such huge numbers, that some scientists believed there were other factors at work. It turned out that zebra mussels, which had inflicted so much ecological pain after invading Lake Erie in 1988, improved conditions for the immature mayflies—called nymphs— that lived in the lake's muddy bottom. The mussels sucked huge quantities of plankton down to the lake bottom, providing a feast for nymphs living in their tiny underwater caves. No one ever proved that zebra mussels were directly responsible for the dramatic resurgence of mayflies. But their recovery four years after zebra mussels were discovered in Lake Erie seemed, to some scientists, like more than a coincidence. Mayfly experts Don Schlosser and Kenneth Krieger explored the factors that contributed to the insects' recovery in a 2000 research report. Their findings were inconclusive but suggestive:

> We do not know the cause for the sudden recolonization of nymphs in large portions of western Lake Erie. Undoubtedly, pollution abatement programs contributed to improved conditions that would have ultimately led to mayfly recovery in the future. However, the explosive growth of the exotic zebra mussel, *Dreissena polymorpha,* undoubtedly diverted plankton foods to bottom substrates which could have increased the speed at which *Hexagenia* nymphs recolonized sediments in western Lake Erie in the 1990s.[2]

That despised zebra mussels contributed to the recovery of mayflies was strangely ironic. It suggested that some native species benefited from the plague of invaders that stormed the Great Lakes in the 50 years after the St. Lawrence Seaway opened. The situation was an inconvenient truth for scientists and environmental advocates who were loath to attribute any ecological benefits to invasive species that colonized the lakes. Would the lakes be better off without zebra and quagga mussels, round gobies and Eurasian ruffe? Without question. But the fact that some native species capitalized on the invaders was a glimmer of hope through the storm of invaders that had conquered the lakes. The lake ecosystems were demonstrating tremendous resilience and adaptability.

Lake Erie's Put-in-Bay has long been one of the most popular vacation destinations in all the Great Lakes. The small bay tucked between South Bass and Gibraltar islands is a place where young children and soccer moms, buttoned-down professionals and college students, jocks and scholars revel in the joys of Midwest summers. South Bass Island is ground zero for Lake Erie's summer party and vacation scene. The island and its surrounding waters are a playground for people whose idea of paradise is a place where they can fish, sail, camp, play golf, track birds or snakes, kayak along the jagged shoreline, or kick back on a boat with a cold drink. For all its human pleasures, and there are many, the island is equally significant for its role in a high-stakes biological drama scripted by invasive species. That was the sole reason I traveled to Put-in-Bay on a June day in 2007. My indulgence in the area's notorious party scene would have to wait—I was headed to a research laboratory that documented the collapse, the resurrection, and great invasion of Lake Erie's ecosystem.

The Jet Express ferry that whisked me and a few other tourists from Port Clinton, on Ohio's mainland, to South Bass Island completed the 12-mile journey in less than a half hour. Once on the island, I walked around the edge of a cove choked with aquatic plants and boarded a water taxi that transported me across the small bay to Ohio State University's Stone Laboratory, on nearby Gibraltar Island. There, I met with Jeffrey Reutter, the lab director and head of Ohio's Sea Grant program. Reutter was a middle-aged scientist and professor who looked like he had just stepped off a hiking trail. Dressed in khaki shorts, a golf shirt, baseball hat, and running shoes, his appearance and engaging personality belied his encyclopedic knowledge of how invasive species affected the Great Lakes—particularly Lake Erie. It was well documented that zebra mussels, quagga mussels, and round gobies transformed Lake Erie's ecosystem, mostly for the worse, and cut short one of the world's greatest examples of ecosystem recovery. To my surprise, the plague of invasives that infected the lakes did cause a few positive changes. I encountered two of those rare success stories while at Gibraltar Island.

I was walking up the boat ramp toward the Stone Laboratory's main building when I noticed something unusual about the building's brick veneer—it seemed to be in motion. Thousands of small tan objects on the building were moving in unison with the wind. The scene resembled a scaled-down version of a field of grain swaying to invisible currents of wind. As I neared the lab, there was a crunching sound underfoot. A noise reminiscent of someone popping bubble wrap was the sound of my shoes crushing mayflies. Hoping to spare as many of the insects as possible, I stepped off the sidewalk and onto the adjacent lawn. That was a mistake. A cloud of mayflies erupted from the grass and enveloped me. The inch-long bugs with oversized transparent wings swarmed my clothes, hair, and seemingly every area of exposed skin. Some tried to sneak into my nose and mouth; others went down my shirt. I covered my nose and mouth with one hand and waved my other arm wildly in a desperate attempt to keep the insects at bay. It was hopeless. The only way to escape the insects—known formally as *Hexagenia,* but commonly called June bugs, Canadian sailors, or shad flies—was to go indoors. Annoying as it was, the cloud of mayflies was welcome news.

Mayflies emerge from Lake Erie as nymphs in early June after spending two years burrowing in the lake's muddy bottom. The nymphs shed their exoskeletons and take flight for what amounts to a three-day insect orgy. Clouds of mayflies large enough to be detected on weather radar mate while flying over the lake. After mating, females release up to 8,000 fertilized eggs on the lake's surface before falling to their deaths. Those eggs not eaten by fish sink to the bottom. There, a new generation of nymphs burrows into the lake bottom before emerging two years later to take flight, shed their skins, and engage in the mating ritual. The process is spectacular, despite the hassles it causes for shoreline communities forced to endure mayfly mayhem for a few weeks each summer. Now, the swarming insects were a sign Lake Erie was healthy again. Mayflies were a keystone species, a barometer of water quality and ecosystem health. What did it mean when millions of mayflies took flight and turned shoreline

Mayflies cling to a window on a building at Ohio State University's Stone Laboratory, located on Lake Erie's Gibraltar Island. The return of mayflies in the 1990s was a sign that Lake Erie's ecological health had improved.

communities into giant insect breeding grounds for weeks at a time? "They're telling us the water's clean out there," said Thomas Edsall, a U.S. Geological Survey biologist.[3]

Mayflies weren't the only native species to flourish in the presence of invasive species. Zebra and quagga mussels that contributed to the collapse of Lake Huron's food chain in 2004 contributed to the decimation of another invader: alewives. That, in turn, allowed two native fish species—walleye and lake trout—to mount long-awaited recoveries. In the absence of alewives, one of Lake Huron's most valuable baitfish species—emerald shiners—returned in

The Lake Erie water snake was one of few native Great Lakes species to benefit from foreign species that invaded the lakes. The once-endangered snake made a dramatic recovery after round gobies infested Lake Erie around 1990. The snakes feast on the foreign fish.

Source: "The Elusive Lake Erie Water Snake" © Benny Mazur, (CC BY 2.0) at: http://www.flickr.com/photos/44545509@N00/2549439764.

huge numbers. And in Lake Erie, the round-goby invasion helped fuel the dramatic recovery of a scorned, but native, snake species.

The Lake Erie water snake is, for my money, one of the vilest creatures native to the Great Lakes. It is also one of the most valuable and symbolic. The snake, *Nerodia sipedon insularum,* resides exclusively along the shores of islands that stretch across Erie's south side. It isn't found anywhere else on the planet. Just because the snake has a penchant for swarming beaches and boat docks, urinating and defecating on anyone who dares to move them, does not diminish its inherent ecological value. Is the Lake Erie water snake less valuable than, say, a bald eagle? Not in the eyes of government officials who declared the water snake a threatened and endangered species in 1999. Decades of abuse by humans, who refused to share with the snakes, had driven the creatures to the brink of extinction.

"People killed the snakes because they wanted to build cottages on the islands and didn't want the snakes around," said Kristin Stanford, a Northern Illinois University graduate student who helped foster the water snake's recovery. "I've had people tell me they would find balls of snakes, pour gas on them and set them on fire. Other people shot the snakes with 22-caliber rifles as they swam in the water."

The light brown snakes, which grow to three-and-a-half-feet long, have been persecuted for years by island residents who don't want them nesting on beaches and in rocky structures used to support boat docks. Part of the hatred is due to the fact that water snakes gather by the hundreds in rocky coastal areas. The snake's ill-tempered behavior has solidified its bad reputation. The creatures aren't poisonous. Hostile, yes. Aggressive? Most definitely. The snakes are notorious for warding off aggressors by emitting a noxious musk that can drive a grown human to the brink of vomiting. If that doesn't work, the animals urinate and defecate on their assailants. Their final line of defense is to bite, which is harmless but still hurts like heck.

Despite the snake's nasty disposition and disgusting defense mechanisms, Stanford fell for the scorned water animal. She was a budding, idealistic biologist when she first traveled to the Stone Laboratory in the summer of 2000. Hers was a singular mission: to save a species reviled by residents of Lake Erie's islands. As other college students partied at Put-in-Bay, Stanford was spending long days collecting and studying water snakes. Seven years later, she was credited with saving the species and given the title of Snake Lady. Stanford had a powerful ally in her snake-resurrection campaign. Water snakes, it turned out, feasted on round gobies, the reviled fish from Europe. With an estimated 10 billion gobies in Lake Erie, many of which lived in the rocky areas around the Bass Islands, the water snake was suddenly living large. The snakes began feasting on gobies and eventually stopped hunting their usual prey: native log perch, stone cats, sculpin, and mud puppies. To water snakes, the gobies were a delicacy. "The snakes started growing faster and larger and the females produced 20 percent more offspring," Stanford said. "The gobies helped bring these snakes back from the brink of extinction."

The water-snake population began to rebound in 2003. By 2007, there were about 7,000 water snakes living around the Bass Islands—well above the goal of 5,000 snakes. As their numbers soared, so did tension among some island residents who were forced to share beaches and boat docks with hostile snakes. "We're getting to the point where there could be another public backlash against the snakes," Stanford said in 2007. "We can find hundreds of snakes at one dock." Barring an unforeseen change, the water snake could be taken off the endangered species list by 2010. "It would be a fabulous success story," Stanford said.[6]

Stanford's work turned her into something of a celebrity. She was featured on television shows and in newspaper and magazine articles. An information kiosk placed on South Bass Island, with a photo of Stanford holding a bunch of snakes in her bare hands, prompted hundreds of e-mails. A West Virginia man sent an e-mail asking for Stanford's snake-bitten hand in marriage. She politely declined (she was already engaged to be married). Stanford was bewildered by all the interest in her unconventional line of work. "It's just weird," she said.

Still, she was happy to share the story of a beleaguered creature returning from the brink of extinction. The snake's recovery was a symbol of hope.

The water snake's unexpected population explosion also was a conundrum for scientists. It was a success story, without question. But the snake's resurgence was largely due to the arrival of round gobies, an invader that terrorized some native fish species and contributed to the deaths of thousands of fish-eating birds. The water snake's recovery, like that of mayflies and walleyes in Lake Huron, was aided by foreign species that disrupted and destabilized Great Lakes ecosystems. Was it possible that gobies were good for Lake Erie? How about zebra and quagga mussels, which dramatically increased water clarity and created conditions that allowed some endangered species of aquatic plants to flourish in the waters surrounding the Bass Islands? Could the foreign species, which were universally scorned by scientists, improve some aspects of Great Lakes ecosystems? I posed the question to Reutter, a distinguished and widely respected Great Lakes scientist with 30 years of research under his belt. His response came quickly: "When you put all of the effects of these invasive species together," he said, "the negative impacts clearly outweigh the positives."[8]

Invasive species plunged the lakes into a state of biological chaos and created great uncertainty about their ecological future. No one knew how the lakes would react, in the long run, to invaders from distant waters. And the few benefits realized by native species that were able to exploit the invaders came with huge caveats. The fate of the Lake Erie water snake, for instance, could be determined largely by two factors: the availability of food and the willingness of humans to share space with the creatures. If the goby population crashed, water snakes might suffer as well.

Similar uncertainty surrounded the mayfly resurrection in Lake Erie. The foreign mussels that bolstered the mayfly's recovery also clouded its future. Zebra and quagga mussels were suspected of contributing to rising concentrations of soluble reactive phosphorus in Lake Erie. The phosphorus was fertilizer for aquatic vegetation and algae, which contributed to the return of the lake's infamous dead zone. If the mussels were indeed responsible for Erie's latest dead zone, mayflies could suffer in the oxygen-starved waters. Round gobies also were a threat to mayflies—the invaders gorged on mayfly nymphs. The complex dynamics left scientists wondering whether gobies, zebra mussels, and quagga mussels would conspire to diminish the mayfly population. It was a sobering question with no answer in 2008, two decades after the mussels turned up in the Great Lakes.

Having written about invasive species in the Great Lakes for two decades, and the failure of government agencies to slow the great invasion, it was hard for me to be optimistic about the future of these imperiled waters. Still, there were major developments in the first decade of the twenty-first century that gave me a genuine sense of hope. Government agencies that had reacted meekly to the problem were finally taking the steps necessary to close the door on new shipborne invaders. The International Maritime Organization had proposed the world's first ballast water treatment standards, and the U.S. Congress was trying to do the same as I completed this narrative in late 2008. Biologists and economists had produced solid data that forced government officials to discuss the possibility of banning ocean freighters from the lakes if shippers could not clean up their vessels' ballast tanks. On a larger scale, the great invasion of the lakes generated a global discussion on the conveyance of species around the planet by ships and other modes of transportation.

Invasive species weren't part of the public discourse in the United States and Canada before zebra mussels turned up in Lake Erie and demonstrated how destructive invasive species could be to aquatic ecosystems and water-dependent economies. Some researchers, in a desperate effort to find good news in the storm of invasives that reconfigured the Great Lakes ecosystem, noted that zebra mussels put the issue of invasive species on the global map. In my humble estimation, the overriding positive that emerged from the plague of foreign species was the public's improved understanding of how fragile the Great Lakes were, despite their enormity and seeming invincibility. Unfortunately, it took three decades of government negligence—and the resulting biological siege—to make that point forcefully enough to bring about the changes needed to protect the lakes from shipborne invaders.

The United States and Canada in 2008 were finally requiring all ocean freighters to flush ballast tanks with seawater before entering the St. Lawrence Seaway and the Great Lakes. That was a positive step, but it was far from a panacea. Ballast water flushing did not eradicate all organisms lurking in the bowels of ocean ships. Potentially deadly pathogens could survive saltwater flushing. That alone was reason enough to require every ship using the St. Lawrence Seaway to install ballast water treatment systems. Sure, treatment systems were expensive, perhaps as much as $1.5 million per ship. But it would be money well spent. Shipping companies that couldn't afford to disinfect their vessels' ballast tanks didn't deserve the privilege of conducting business on the lakes.

The global shipping industry was not entitled to transport cargo on the Great Lakes. Use of the world's largest freshwater reservoir was a privilege afforded the industry when the United States and Canada built the St. Lawrence Seaway. Clearly, the industry never should have been allowed to dump untreated ballast water into the lakes after U.S. lawmakers passed the federal Clean Water Act in 1972. Because the lakes are a closed ecological system, any chemical or biological pollutants dumped in them will likely remain for decades, if not centuries. That reality made it all the more important to treat these freshwater seas with the utmost care.

A 2008 study by the U.S. National Research Council concluded that the only sure way to keep ships from importing more foreign species into the lakes was to close the St. Lawrence Seaway to all freighters. But the council, an advisory panel to the U.S. Congress, recommended against it. Such a move, the council concluded, would eliminate the potential for future increased global trade in the Great Lakes region. Essentially, the council told Congress that if the United States and Canada wanted international trade to continue on the lakes, the two nations had to accept the risk that transoceanic ships might import more foreign species. Rather than close the Seaway, the council urged the U.S. Congress to immediately adopt ballast water treatment standards proposed by the International Maritime Organization. That was all fine and good, but the fact remained that no ballast water treatment system ready for use in 2009 was 100 percent effective. Moreover, it would take several years to equip all ocean freighters that visited the Great Lakes with ballast treatment systems. Until that day arrived, the lakes remained vulnerable to more shipborne invasions. In the interim, residents of the region could only hope that foreign ships wouldn't deposit more destructive invaders in the lakes.

If we've learned nothing else from the ecological ruin inflicted by shipborne invaders, we should now understand that in the war against invasive species, prevention is the best—perhaps the only—guaranteed protection. No public health official would condone unprotected sex between a person with herpes and an uninfected individual. Why, then, would we allow ships carrying cholera and other pathogens in ballast tanks to engage in unprotected trade in the Great Lakes? It defies logic. What if a transoceanic ship imports another foreign organism that causes billions of dollars of additional ecological damage? What if a ship's ballast water discharge infects a public drinking water system with cholera or another deadly pathogen that kills humans? Would that trigger the kind of dramatic, rapid response needed to end the inoculation of the lakes by shipborne invasive species? Those were the questions that rang in my mind as I headed back to mainland Ohio.

The sun was sinking toward the Lake Erie horizon when I boarded the Jet Express ferry in Put-in-Bay and headed back to the more tranquil community of Port Clinton, Ohio. The steady hum of the ferry's engines and the melodic rhythm of the ship carving through Erie's mild chop set my mind adrift. Passing anglers in small fishing boats that bobbed in the expansive lake, I wondered if future generations would come to know these emerald waters as the Walleye Capital of the World.

After disembarking from the ferry, I hopped in my car and drove a few miles east before settling on a hotel along Lake Erie' of mayflies greeted me as I stepped out of my car. Thousands swarmed around the hotel's exterior lights, while others snapped underfoot as I scurried into the building. I was moving quickly in a desperate attempt to get checked in and deposit my luggage in the hotel room before the sun set. There were few better places than Great Lakes beaches to watch the sun illuminate the evening sky as it sank below the horizon. Lake Erie did not disappoint that night. Moisture that evaporated from the lake formed a thin layer of cirrus clouds that resembled giant horse manes. The clouds provided the canvas upon which the sun cast a glorious pale-red hue as it neared the horizon. Hard as I tried to enjoy the synergistic beauty of the lake, sky, and sun, my thoughts were haunted by an ominous sense of uncertainty.

Would children not yet born in 2009 fall in love with the lakes by playing in clean waters that gently lapped at unspoiled shorelines? Would they come to know the lakes as an intricately woven ecosystem that supported tiny freshwater shrimp, majestic bald eagles, giant lake sturgeon, and delectable whitefish? Would they be able to rely on the lakes as a source of clean drinking water? Would they view the lakes as a priceless resource that warranted protection at any cost? Or would they come to know the lakes as victims of a biological cleansing carried out by an army of invasive species from distant waters? Would they be repulsed by algae blooms that fouled the water, harbored potentially deadly organisms, and turned beaches into avian killing fields? Would they smell the musty odor of blue-green algae when they drew a glass of drinking water from the tap? Would they come to know the simple, profound joy of hearing a common loon cast its tremulous call across the surface of a Great Lakes bay on a tranquil summer morning? In this, the 50th anniversary of the St. Lawrence Seaway's opening, there was no way of knowing what the future held for the lakes. There were many vexing questions but few answers.

Foreign species that transoceanic freighters accidentally deposited in the lakes had caused the most profound changes to the lakes in recorded history. Scientists could not envision how the lakes' ecosystems would function in coming decades—they couldn't even keep pace with profound changes that unfolded before their eyes at the dawn of the twenty-first century. Granted, the lakes had proven remarkably resilient in the past. These freshwater seas—Superior, Michigan, Huron, Erie, and Ontario—staged magnificent ecological recoveries in the 1970s after humans eased up on the gas peddle of exploitation and abuse. But the conquest of the lakes by foreign species was different. There was no chemical, insect, or animal capable of controlling the range or densities of zebra and quagga mussels, round gobies or Eurasian ruffe, bloody red shrimp or spiny water fleas in the lakes. Humans knew how to split atoms and build spacecraft that allowed astronauts to live beyond the reach of Earth's gravity. Yet, once foreign species stormed the lakes, we were largely powerless against the army of invaders that crippled native species, spawned toxic algae blooms, and formed bastardized food chains that delivered toxic chemicals and deadly pathogens to fish, wildlife, and humans.

Looking across Lake Erie on that sultry summer evening, I was struck by the contrast of what my eyes could see and what my mind knew to be true. The shimmering surface of the lake was alluring, exquisite. But its beauty was skin deep. Beneath its winsome veneer, an ecological war raged and the bad guys were winning big. As the sun dipped below the horizon and the sky grew dark, I wondered if Lake Erie—the smallest, shallowest, most productive, and most fragile of all the Great Lakes—would weather the storm of invasive species. Erie's fate was enormously significant. The lake was an ecological guinea pig, a bellwether of future changes in Lakes Ontario, Huron, Michigan, and Superior. That Erie was losing the battle against invasive species was cause for alarm. But its remarkable history provided a glimmer of hope for the beleaguered lakes. Lake Erie, after all, recovered magnificently after it was prematurely declared dead in the 1960s.

Perhaps Erie, along with the other Great Lakes, would survive the reign of ecological terror carried out by invasives and reclaim some of their natural splendor. We could hope for such a turn of events, much like cancer patients who dream of restored health. The alternative emotion, despair, was unworthy of these inimitable waters. There was no changing the fact that the Seaway, and all it allowed into the heart of North America, stripped the lakes of their natural defenses and sickened these freshwater seas. All that remained in 2009 for the Great Lakes, and all whose lives were intricately tied to these priceless waters, were the possibilities that hope afforded. We could hope for a brighter future for the lakes and ourselves—just like the mythical Pandora.

FINAL NOTES

Haudenosaunee (Iroquois) Great Law of Peace, Part III

By Ellis E. McDowell-Loudan

In *Human Evolution and Survival,* I have tried to inject the importance of finding out more about ourselves as part of the natural environment and the remarkable ongoing-changing world around us. Instead of a traditional textbook, the articles provided are embellishments and examples to broaden the scope. They explain why and how cultural baggage has continued to plague us, despite efforts to update and upgrade education.

When new discoveries are made, sometimes they make older ones obsolete. Other times, they explain inconsistencies. The multiple traits present in our genetic makeup and the way in which we understand them today are quite a change from previous views. The new fossil finds often turn out to be older than the previous "oldest" ones, providing new "dawns of hominins."

The Neandertal controversies and continuing discoveries, with remarkable developments which add cultural-archaeological information and biogenetic information through DNA, are thrilling in their variety. Who knew, a mere decade ago, that there could be so much more evidence and geographic diversity?

Ethical issues, no matter what the discipline—Anthropology, Biology, History, Political Science, or any other—and no matter what time period continue to be of significance. The nature of them, the variation, and some issues are characteristic of today, however. In some instances, they correlate well with the other topics in the text. Careers, issues of human diversity, class, age, gender, environmental devastation—all correlate with the topic of ethics in one way or another.

A problem with confronting some of these topics is that there continue to be major differences of opinion as to the nature of human rights. Even in our own society, there are diverse perspectives when the topic arises, and the ways and means of addressing it are considered.

It is with this in mind that I decided to end my book with a quote from Tom Porter (Sakokwenionkwas), Bear Clan Elder of the Mohawk Nation of the Haudenosaunee. It is excerpted from Chapter 28, "The Great Law of Peace Part 3: Keeping and Preserving It," page 313:

> "Put your whole heart in what you learned, so that the spirit of our ancestors, our Creator and our Peacemaker can talk right through your body and use you, so that you can communicate peace. And then our people, we will get better. Our people will be united again."
> [*And Grandma Said ... Iroquois Teachings as Passed Down through the Oral Tradition.* Tom Porter; transcribed and edited by Lesley Forrester, 2008. Xlibris Corporation.]

While this statement refers to the Haudenosaunee or Iroquois as spoken and transcribed in Porter's book, the message, in my view, is one that has value. What we learn—the insights we gain from parents, from our life experiences, including schools, should be gathered wholeheartedly, so that the patterned learned behavior of our lives will be transmitted in the best of all possible ways to the next generation! Each small step is a valuable one—the fingerprinting project in India, which ideally will improve health care conditions for all through identification, is one example. The publication and sharing of Tom Porter's book for use throughout the Six Nations of the Haudenosaunee, as well as in many colleges and universities for Native American Studies Programs, provide another opportunity for sharing of this wisdom.

> "We walk in the footsteps of our ancestors. We will be known for the tracks we leave." Doug Sky Jones, Onondaga Nation

Tom Huff, Cayuga-Seneca Nations, soapstone carving